Lecture Notes in Chemistry

Volume 106

The series Lecture Notes in Chemistry (LNC), reports new developments in chemistry and molecular science - quickly and informally, but with a high quality and the explicit aim to summarize and communicate current knowledge for teaching and training purposes. Books published in this series are conceived as bridging material between advanced graduate textbooks and the forefront of research. They will serve the following purposes:

- provide an accessible introduction to the field to postgraduate students and nonspecialist researchers from related areas,
- provide a source of advanced teaching material for specialized seminars, courses and schools, and
- be readily accessible in print and online.

The series covers all established fields of chemistry such as analytical chemistry, organic chemistry, inorganic chemistry, physical chemistry including electrochemistry, theoretical and computational chemistry, industrial chemistry, and catalysis. It is also a particularly suitable forum for volumes addressing the interfaces of chemistry with other disciplines, such as biology, medicine, physics, engineering, materials science including polymer and nanoscience, or earth and environmental science.

Both authored and edited volumes will be considered for publication. Edited volumes should however consist of a very limited number of contributions only. Proceedings will not be considered for LNC.

The year 2010 marks the relaunch of LNC.

More information about this series at http://www.springer.com/series/632

Franca Morazzoni

Symmetry in Inorganic and Coordination Compounds

A Student's Guide to Understanding Electronic Structure

Springer

Franca Morazzoni
Department of Materials Science
University of Milano-Bicocca
Milan, Italy

Chapters 1–4 contain text from Morazzoni, F: Symmetry in Inorganic and Coordination Compounds—A Student's Guide to Understanding Electronic Structure, Universitas Studiorum, Mantova, Italy (2018), that has been reprinted with the permission of Universitas Studiorum S.r.l.

ISSN 0342-4901 ISSN 2192-6603 (electronic)
Lecture Notes in Chemistry
ISBN 978-3-030-62995-3 ISBN 978-3-030-62996-0 (eBook)
https://doi.org/10.1007/978-3-030-62996-0

This Springer imprint is published by the registered company Springer Nature Switzerland AG
The registered company address is: Gewerbestrasse 11, 6330 Cham, Switzerland

Preface

The collected lectures summarize the contents of the course "Physical Methods in Inorganic Chemistry." The first time I gave this course has been in 1981 at the University of Milano and then from 2000 to 2017 at the University of Milano-Bicocca, always dedicated to the advanced level students of chemistry and materials science.

The reader is here conducted, by the symmetry properties of the molecules, to use the chemical bond models suitable to explain the properties of inorganic and coordination compounds.

Special emphasis was given to the crystal field theory and to the possibility of designing the energy level trend by using the electronic spectra and the magnetic measurements. In 1981, this constituted a frontier research topic, which then became matter to be regularly taught.

When I am preparing the course the first time, my daughter Diletta was 3 years and she asked me each evening to repeat the lecture of the day after, sat on her small bed. In two years, she was able and want to directly write her name, on the transparencies at that time used during the lectures. I dedicate this book her, who is now a young women, is an economist and fortunately she did not decide to follow the mother science traces.

Milan, Italy

Franca Morazzoni

Acknowledgements

I want to thank all the colleagues which helped me to follow the chemistry pathway of the symmetry rules and appear as coauthors in most of the cited publications—first of all, my Ph.D. supervisor, the late Professor Franco Cariati, pioneer of the crystal field studies, and then the researchers of my group of the University of Milano-Bicocca Roberto Scotti, Massimiliano D'Arienzo and Barbara Di Credico, which strongly contributed to make possible the case studies reported in this book.

Introduction

The knowledge of the molecular nature of matter and of its changes constitutes at the present the necessary pathway to understand the chemical properties and the reactivity. The goal is, based on molecular properties, also inclusive of the electronic properties, to reach more valuable and extensive laws in chemistry. The complexity of studying even a simple inorganic molecule suggests that approximate methods have to be used in the quantum mechanical treatment, taking also into account the macroscopic peculiarities of the chemicals, e.g., their molecular shape.

In the case of inorganic molecules, a convenient approximation source comes from the symmetry, which constrains the electronic energies and the chemical bonds to follow its rules. Thus, the first part of the present text will give special emphasis to the symmetry properties in the frame of group theory and will compare the use of symmetry operators with that of Hamiltonian operators. If possible, also the reactivity of the molecules will be rationalized in terms of symmetry properties. Electronic spectroscopy and magnetism will be used to provide the experimental confirmation of the electronic energy levels.

The description starts with a short summary of the variation and the perturbation methods together with a limited description of the group theory, in order to demonstrate how the symmetry rules simplify the application of the methods.

The electronic structure of simple inorganic molecules is then described as an example by symmetry-adapted linear combinations of atomic orbitals.

Well-studied coordination compounds with different symmetries, regular and distorted, are described by using the crystal and ligand field approaches.

Their electronic structure is confirmed by electronic spectra and magnetic measurements.

Finally, a large number of examples taken from the scientific literature, reported the capability of the so determined electronic structure to explain the chemical properties of the systems and in some case their functionality (both chemical and physical).

An appendix of mathematics useful to understand the theoretical descriptions is reported.

The book is a collection of lectures, and thus, it cannot be considered exhaustive for the preparation of the specific exam; however, it constitutes a useful guideline for each chemist to approach the physical-chemistry properties of inorganic and coordination compounds looking to their functionality.

Contents

About the Author

Prof. Franca Morazzoni, Ph.D. in Chemistry at the University of Milano, became Associated Professor and then Full Professor of Inorganic Chemistry at the University of Milano-Bicocca. She funded the group of hybrid nanomaterials at the Department of Material Science in 2001. As an expert of physical methods in inorganic chemistry applied to hybrid materials, she published more than 250 paper in international journals. She was responsible of several national research projects in the field, participated to the development of inorganic nanofillers with controlled shape, to be included in tires, in collaboration with Pirelli Tyre. Several patents were signed on this topic. She was Vice Chair of the HINT COST action (2012–2016) on the organic-inorganic hybrid interface. Professor Morazzoni has supervised several Ph.D. students and until 2015 was director of the Ph.D. School of the University of Milano-Bicocca. She plays piano giving regularly performance.

Chapter 1
The Electronic Structure Determination

This is a peculiar physical problem, and only later in the time, compared with the atomic structure determination by physical scientists, the chemists investigated the molecular case study. Expectedly during the time, the preferential object for chemists has been focused on the rationale of the reactivity, where a number of experimental data are available. The electronic structure of the molecules, mainly of inorganic ones, slowly began to be investigated.

Starting from this situation and in order to approach the electronic structure determination, the following pathway is proposed:

(i) The Schrodinger equation $H\Psi_n = E\Psi_n$ is very difficult to be resolved for a molecular system, being the E_n values

$$E_n = \langle \Psi_n^* | H | \Psi_n \rangle / \langle \Psi_n^* | \Psi_n \rangle$$

impossible to be calculated in the absence of H and Ψ_n expressions.

(ii) Hypotheses on Ψ_n and/or H have to be made in order to optimize Ψ_n by *variation* or *perturbation* methods.

(iii) The eigenvalues of the Schrödinger equation are calculated by assuming approximated values of the integral $\langle \Psi_n^* | H | \Psi_n \rangle$, in agreement with the peculiar properties of the compound.

(iv) The calculated eigenvalues are compared with the experimental ones, derived from electronic spectra, in order to obtain structure–chemical property relations.

F. Morazzoni, *Symmetry in Inorganic and Coordination Compounds*, Lecture Notes in Chemistry 106, https://doi.org/10.1007/978-3-030-62996-0_1

1.1 Variation Method (Molecular Orbital Linear Combination of Atomic Orbitals, MO-LCAO)

By using the variation method, the hypothesis on Ψ_n is that it consists in a linear combinations of atomic orbitals (AOs)

$$\Psi_n = \Sigma_r c_{nr} \varphi_r$$

r indicates the rth atom
The hypothesis on H is absent
For a biatomic bielectronic system,

$$\Psi = C_1\varphi_1 + C_2\varphi_2 \quad E = E(C_1 C_2)$$

$$E = \int \Psi^* H \Psi d\tau / \Psi^* \Psi d\tau$$

$$= \int (C_1\varphi_1 + C_2\varphi_2)^* H(C_1\varphi_1 + C_2\varphi_2) d\tau / (C_1\varphi_1 + C_2\varphi_2)^* (C_1\varphi_1 + C_2\varphi_2) d\tau$$

$$= \frac{\int [C_1\,\varphi_1{}^* HC_1\varphi_1 + C_1\,\varphi_1{}^* HC_2\varphi_2 + C_2\,\varphi_2{}^* HC_1\varphi_1 + C_2\,\varphi_2{}^* HC_2\varphi_2]/}{[C_1\,\varphi_1{}^*\, C_1\varphi_1 + C_1\,\varphi_1{}^*\, C_2\varphi_2 + C_2\,\varphi_2{}^*\, C_1\varphi_1 + C_2\,\varphi_2{}^*\, C_2\varphi_2]}$$

if $\int \varphi_i{}^* H \varphi_j d\tau = H_{ij}$ and $\int \varphi_i{}^* \varphi_j d\tau = S_{ij}$ $\quad H_{21} = H_{12}\ S_{21} = S_{12}$

H_{ii} = Coulomb integral
H_{ij} = resonance integral
S_{ij} = overlap integral

$$E = C_1{}^2 H_{11} + 2C_1C_2H_{12} + C_2{}^2 H_{22} / C_1{}^2 S_{11} + 2C_1\, C_2{}^2 S_{12} + C_2{}^2 S_{22}$$

The energies have to be minimum varying C_1 and C_2 (variation condition)

$$\begin{aligned} dE/dc_1 &= 0 \\ dE/dc_2 &= 0 \end{aligned} \tag{1.A}$$

(1.A) is a two equation linear system, to derive $E_{1,2}$ and $C_{1,2}$ values.

$E_{1,2}$ are depending on the Coulomb, and on the resonance and overlap integrals, which are difficult to be calculated. The system (1.A) becomes

$$\begin{aligned} C_1(H_{11} - E\,S_{11}) + C_2(H_{12} - E\,S_{12}) = 0 \\ C_1(H_{21} - E\,S_{21}) + C_2(H_{22} - E\,S_{22}) = 0 \end{aligned}$$

One solution is $C_1 = C_2 = 0 \quad \Psi = 0$

The other ones may be obtained from the following determinant:

$$\begin{vmatrix} H_{11} - ES_{11} & H_{12} - ES_{12} \\ H_{21} - ES_{21} & H_{22} - ES_{22} \end{vmatrix} = 0$$

If the two atoms are the same

$$(H_{12} - ES_{11})^2 - (H_{12} - ES_{12})^2$$

Given $S_{11} = S_{22} = 1 \quad S_{12} = S_{21} \quad H_{11} = H_{22} \quad H_{12} = H_{21}$

$$E_1 = H_{11} + H_{12}/1 + S_{12}$$
$$E_2 = H_{11} - H_{12}/1 - S_{12}$$

The determinant dimension is equal to the number of atomic orbitals used in the linear combination. The E energy values substituted in (1.A) and combined with the additional equation $C_1^2 + C_2^2 = 1$ allow to calculate the C coefficients.

S_{ij} is obtained from the molecular structure.

The integrals H_{ij} cannot be precisely determined for complex molecules, and all methods denominated MO-LCAO differ in the different approximations to calculate the H_{ij} values. The calculations of both energies and the atomic orbital coefficients can be done by solving an eigenvalues and eigenvectors equation.

The most simple approximation is the Hückel approximation essentially based on symmetry properties and frequently used for organic molecules.

1.2 A Matrix Formulation of MO Calculations

The solution of an eigenvalue problem can be obtained by a simple matrix formulation.

The solution of the wave equation is eigenvalues of the Hamiltonian operator; thus, $H\,\Psi = E\Psi$ can be written in matrix form as

$$[H_{ij}][C] = [S][C][E]$$

[C] is the matrix coefficient, $[H_{ij}]$ is the energy matrix of the elements $\langle \varphi_i | H | \varphi_j \rangle$, [$S$] is the overlap matrix, and [E] is the diagonal matrix of the orbital energies.

Dividing both sides by [C]

$$[C]^{-1}[H_{ij}][C] = [S][E]$$

The energies result from the diagonalized $[H_{ij}]$ matrix as well as the LCAO coefficients are the terms of the diagonalizing matrix.

1.3 The Hückel Approximation

The most simple approximation in the frame of MO-LCAO is the Hückel approximation, essentially based on molecular symmetry properties, and most frequently used for organic molecules.

The secular equations, written directly by generalizing the variational case described above for a two atomic orbital system, become

$$C_{11}(H_{11} - E\,S_{11}) + \cdots C_{1n}(H_{1n} - E\,S_{1n}) = 0$$
$$C_{n1}(H_{n1} - E\,S_{n1}) + \cdots C_{nn}(H_{nn} - E\,S_{nn}) = 0$$

Given the requirements for the approximation:

(1) Resonance integrals = 0 for unconnected atoms, instead
= β (scalar value) for connected ones
(2) Coulomb integrals = α (scalar value)
(3) Overlap integrals = 1 for connected atoms, instead
= 0 for unconnected ones

The secular equations become

$$C_{11}(\alpha - E) + C_{12}\beta_{12}\ldots C_{1n}\beta_{1n} = 0$$
$$C_{n1}\beta\ldots C_{nn}(\alpha - E) = 0$$

The results of calculation, obtained by solving the above linear system, allow to express the energies as a function of α and β that in turn can be determined by using the experimental absorption energy values.

The simple example is reported in the following.

1.3.1 The Case of the Allyl Anion (MO-LCAO by π Orbitals)

The Hückel method allows to obtain the energy of MO π orbitals, constituted by three p atomic orbitals of the three carbon atoms (Scheme 1.1), as eigenvalues of the appropriate secular determinant. At the same time, the coefficients of the atomic orbitals within the molecular orbitals are obtained as eigenvectors of the same determinant.

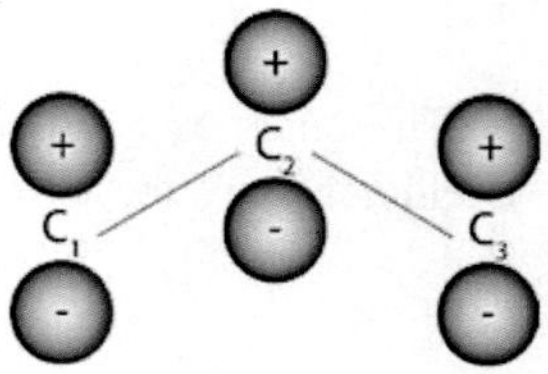

Scheme 1.1 The π system of the allyl radical anion

The secular determinant for this system becomes

$$\begin{matrix} \alpha - E & \beta & 0 \\ \beta & \alpha - E & \beta \\ 0 & \beta & \alpha - E \end{matrix} = 0$$

The zero terms arise from carbon atoms 1 and 3 which are unconnected. Dividing by β and letting (α − E)/β = X, the secular determinant is

$$\begin{matrix} X & 1 & 0 \\ 1 & X & 1 \\ 0 & 1 & X \end{matrix} = 0$$

resolved as

$$X\begin{vmatrix} X & 1 \\ 1 & X \end{vmatrix} - 1\begin{vmatrix} 1 & 1 \\ 0 & X \end{vmatrix} = 0$$

$X^3 - 2X = 0$ with solutions $X = 0$ $X = \pm\sqrt{2}$ that is

$$(\alpha - E)/\beta = 0, +\sqrt{2}, -\sqrt{2}$$

These three roots produce the three energies

$$E_1 = \alpha + \sqrt{2}\beta$$
$$E = \alpha$$
$$E_3 = \alpha - \sqrt{2}\beta$$

The secular equations can be written by taking the general equations and multiplying the column matrix [C_1 C_2 C_3] by the 3 × 3 determinant.

$$C_1X + C_2 = 0$$
$$C_1 + C_2X + C_3 = 0$$
$$C_2 + C_3X = 0$$

When $X = -\sqrt{2}$

$$C_2 = C_2/\sqrt{2} \quad C_3 = C_2/\sqrt{2}$$

Using the normalization criterium

$$C_1^{\,2} + C_2^{\,2} + C_3^{\,2} = 1 \quad 2C_2^{\,2} = 1 \quad C_2 = 1/\sqrt{2} \quad C_1 = C_3 = 1/2$$

When $X = 0 \quad C_2 = 0 \quad C_1 = -C_3 \quad C_1 = 1/\sqrt{2}$
The eigenvalues are thus

$$\Psi_1 = 1/2\varphi_1 + 1/\sqrt{2}\varphi_2 + 1/2\varphi_3 \quad \textit{Bonding orbital}$$
$$\Psi_2 = 1/\sqrt{2}\varphi_1 - 1/\sqrt{2}\varphi_3 \quad \textit{Nonbonding orbital}$$
$$\Psi_3 = 1/2\varphi_1 - 1/\sqrt{2}\varphi_2 + 1/2\varphi_3 \quad \textit{Antibonding orbital}$$

The energy difference between the bonding or antibonding orbital and the non-bonding is $\pm\, \beta \sqrt{2}$; thus, from the assigned experimental transitions, the β value can be calculated. The Hückel approximation is a semiempirical approach. The α value is common intermediate energy level and plays as a reference.

Basing on the results of the Hückel approximation, the electron density on the allyl ion is concentrated on the terminal carbons, in fact:
If we define the electron density on the r atom as

$$Q_r = -\,\Sigma_j n_j\, c_{jr}^{\,2} \quad n_j \text{ is the electron number in the jth MO}$$

c_{jr} is the coefficient of the rth atomic orbital (AO) in the jth MO for the Hückel obtained filled orbitals we obtain $q_1 = -3/2$ $q_2 = -1$ $q_3 = -3/2$
The electronic charge located on the terminal C atoms is higher than the intermediate one; thus, we can expect that the reactions involving a dipolar interaction are oriented in a way as the terminal atoms interact with the positive pole of the reactant, e.g.,

$$CH_3 - CH = CH_2 + R^{-}MgX^{+} \rightarrow CH_2 = CH - CH_2^{-\cdot} + RH + MgX^{+}$$
$$^{-\cdot}CH_2 - CH = CH_2 + MgX^{+} \rightarrow CH_2MgX - CH = CH_2$$

1.4 The Extended Hückel Procedure

This procedure allows to assume for the energy integrals, suitable values taken from the experimental results. In this approximation, for H_{ii} is taken the valence-state ionization potential. H_{ij} is given by the Wolfsberg–Helmholz 1/2k Sij (Hii + Hjj) still obtained from the ionization data.

Alternatively, the Cusach's formula $H_{ij} = 1/2\ (2 - Sij)\ S_{ij}\ (Hii + Hjj)$
Tables 1.1 and 1.2 report the atomic orbital (AO) coefficient calculated by the extended Hückel approximation, for any single molecular orbital (MO), in the allyl ion (Scheme 1.2).

Table 1.1 Extended Huckel coefficients for allyl anion ([1], p. 70)

(The vertical numbers correspond to A O.'s and the horizontal ones to M.O.'s)

A.O.'s ↓ \ M.O.'s →	1	2	3	4	5	6	7	8_n	9_x
1	−1.80	−0.00	0.24	−0.00	0.27	−0.00	0.25	−0.00	0.00
2	0.00	0.00	0.00	0.00	0.00	−0.00	0.00	−0.97	−0.00
3	−0.51	−0.00	−0.35	0.00	−1.31	0.00	−0.24	0.00	−0.00
4	0.00	−1.28	−0.00	−0.68	−0.00	−0.29	−0.00	−0.00	−0.00
5	0.76	1.12	0.87	−0.38	−0.01	0.22	0.01	−0.00	0.00
6	−0.00	−0.00	−0.00	0.00	0.00	−0.00	0.00	0.65	−0.74
7	−0.43	−0.53	0.31	−0.58	0.30	0.62	−0.65	0.00	0.00
8	−0.68	−0.13	0.58	−0.89	−0.53	−0.54	0.15	−0.00	0.00
9	0.76	−1.12	0.87	0.33	−0.01	0.22	0.01	−0.00	−0.00
10	−0.00	−0.00	−0.00	0.00	−0.00	0.00	−0.00	0.65	0.74
11	−0.43	0.53	0.31	0.58	0.30	−0.62	−0.65	−0.00	−0.00
12	0.68	−0.13	−0.58	−0.89	0.53	−0.54	−0.15	0.00	0.00
13	0.28	0.00	−0.30	0.00	−1 17	0.00	−0.61	−0.00	0.00
14	−0.07	−0.48	−0.61	0.49	0.46	0.68	−0 59	0.00	0.00
15	0.15	−0.09	−0.75	0.82	−0.03	−0.55	0.61	−0.00	−0.00
16	0.15	0.09	−0.75	−0.82	−0.03	0.55	0.161	0.00	0.00

		10	**11**	**12_x**	**13**	**14**	**15**	**16**	**17**
	1	−0.00	0.01	−0.00	0.00	−0.09	0.33	−0.00	0.48
C	2	−0.00	−0.00	−0.56	0.00	−0.00	−0.00	−0.00	0.00
	3	−0.00	0.33	−0.00	−0.00	−0.36	−0.17	−0.00	0.01
	4	−0.49	−0.00	−0.00	−0.30	0.00	−0.00	0.16	−0.00
	5	−0.14	0.05	0.00	−0.03	−0.03	−0.20	0.44	0.36
C	6	−0.00	0.00	−0.40	−0.00	0.00	−0.00	0.00	0.00
	7	−0.12	−0.35	−0.00	0.37	−0.12	−0.12	0.01	−0.00
	8	0.41	0.00	−0.00	0.07	0.29	−0.14	0.02	−0.01
	9	0.14	0.05	−0.00	0.03	−0.03	−0.20	−0.44	0.36
C	10	0.00	0.00	−0.40	0.00	−0.00	−0.00	0.00	0.00
	11	0.12	−0.35	−0.00	−0.37	−0.12	−0.12	−0.01	−0.00
	12	0.41	−0.00	−0.00	0.07	−0.29	0.14	0.02	0.01
	13	0.00	−0.40	−0.00	0.00	0.27	0.32	0.00	0.08
H	14	0.37	0.26	0.00	−0.15	0.27	−0.15	0.17	0.04
	15	−0.07	−0.33	0.00	0.36	0.02	−0.23	0.19	0.05
	16	0.07	−0.33	0.00	−0.36	0.02	−0.23	−0.19	0.05
	17	−0.37	0.26	−0.00	0.15	0.27	−0.15	−0.17	0.04

M.O.'s 10–17 have $2e^-$
M.O.'s 9 has le^-
M.O.'s 1–8 empty
Atomic orbitals 1–4, 5–8, and 9–12 are carbon orbitals in the order 2_s, $2p_s$, $2p_x$, $2p_y$, for each atom
A.O.'s 13–17 are hydrogen 1*s*

Table 1.2 Energy of MO in allyl anion ([1], p. 71)

M.O.	Energy (eV)	Occ.
1	53.22	0
2	36.55	0
3	28.31	0
4	22.97	0
5	18.76	0
6	7.30	0
7	5.19	0
8	−4.63	0
9	−10.16	1
10	−12.67	2
11	−13.83	2
12	−14.03	2
13	−15. 01	2
14	−15.96	2
15	−19.40	2
16	−22.83	2

M.O. 12 is the lowest energy occupied molecular orbital having π symmetry
It is composed of $-0.40\ 2p_z(C_2)\ -0.56\ 2p_z(C_1)\ -0.40\ 2p_z(C_3)$ $E = -14.03$ eV. This is in agreement with the results of the simple Hückel approach because in the case of allyl all the Hückel requirements are satisfied

Scheme 1.2 Allyl anion

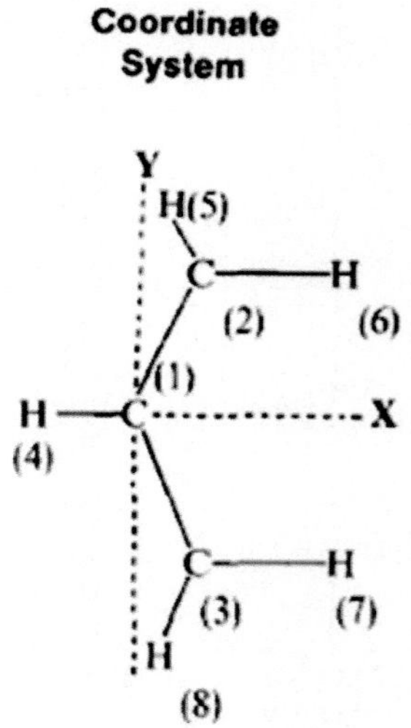

In all cases until now considered, it is mandatory to select the atomic orbitals which have to be used in the linear combination and give the MO. As the maximum limit, all AOs can be employed. However, it is not convenient to consider all the orbitals because some of them give no relevant contribution to MO (see allyl anion Table 1.1 of M.O and A.O.)

How to select those which give the maximum contribution?

Let us consider that the MO shape should necessarily reflect the symmetry of the molecule and the AO has to be combined in agreement with this symmetry. In order to select the symmetry of MO and to satisfy the symmetry imposal, we have to use the *group theory*.

WHAT DOES IT MEAN?

Let us consider firstly the details of symmetry elements and of the related symmetry operations in a molecule (Table 1.3), which are the molecule motions reproducing it unchanged.

As example of operation, it is here reported the clockwise rotation on the BCl_3 molecule and the reflection by a mirror plane for the H_2O molecule. The symbols of the symmetry operations are those reported in Table 1.3.

C_n^n indicates a n-order rotation, repeated n times.

Table 1.3 Symmetry elements and associated operations

Symmetry operation	Symmetry element	Symbol	Examples
Identity		E	All molecules
Reflection	Plane	σ	H_2O, BF_3(planar)
Inversion	Point (center of symmetry)	i	
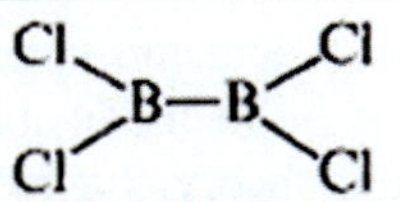			
Proper rotation	Axis	C_n (*n*-order)	NH_3, H_20
Improper[a] rotation (rotation by $2\pi/n$ followed by reflection in plane perpendicular to axis)	Axis *and* plane	S_n (*n*-order)	Ethane, ferrocene (staggered structure)

[a]Ferrocene is staggered and possesses an S_{10} improper rotation axis

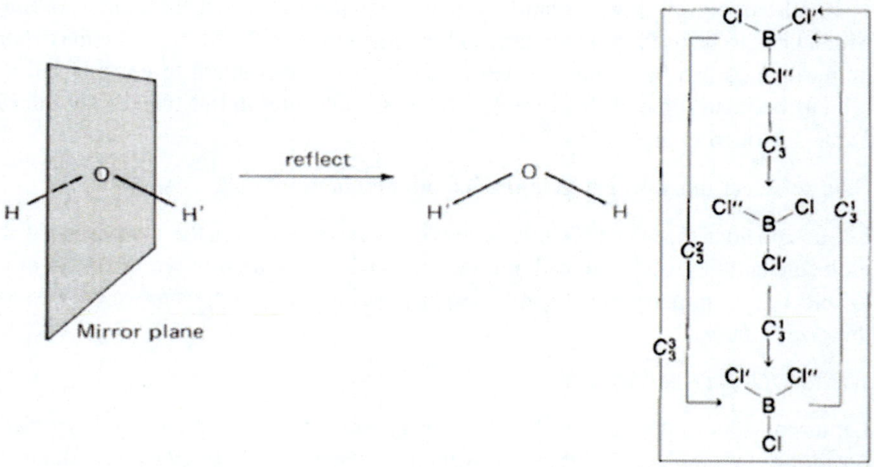

Examples of symmetry operations

Since the molecular energy is time invariant and independent of the position of the molecule, the Hamiltonian must be invariant under a symmetry operation. Thus, any symmetry operation is associated with a symmetry operator which commutes with the Hamiltonian, i.e., $HR = RH$ (R is the symmetry operator associated to the symmetry operation). If we take the wave equation $H\Psi_n = E\Psi_n$ and multiply each side by the symmetry operator R, we get

$HR\Psi_n = ER\Psi_n$; thus, $R\Psi_n$ is a eigenfunction
Being Ψ_n normalized, it is normalized also $R\Psi_n$; that is, $\int R\Psi_n R\Psi_n \, d\tau = 1$ ($d\tau$ is the infinitesimal coordinate space) and

$$R\Psi_n = \pm 1\Psi_n$$

The molecular eigenfunction does not vary under the symmetry operators of the molecular point symmetry group, as well as the base functions of an irreducible representation base function (see further details in the next group theory chapter).

1.5 Group Theory

The point groups collect all the possible symmetry operations for a given molecule. Each symmetry element reported in Table 1.4 can be associated to the symmetry operations.

The *following flowchart* (Scheme 1.3) provides a systematic way to classify the molecules in their proper point groups, through the identification of the symmetry elements.

Table 1.4 Symmetry elements in some common point groups

Point group	Symmetry elements[a]	Examples
C_1	No symmetry	SiBrClFI
C_2	One C_2 axis	H_2O_2
C_{nh}	One n-fold axis and a horizontal plane $\sigma_h$which must be perpendicular to the n-fold axis	*trans*-$C_2H_2Cl_2$ (C_2h)
C_{2v}	One C_2 axis and two σ_v planes	H_2O, SO_2Cl_2, $SiCl_2Br_2$
C_{3v}	One C_3 axis and three σ_v planes	NH_3, CH_3Cl, $POCl_3$
D_{2h}	Three C_2 axes all ⊥ , two σ_v planes, one σ_h plane, and a center of symmetry	N_2O_4 (planar)
D_{3h}	One C_3, three C_2 axes ⊥ to C_3, three σ_v planes, and one σ_h	BCl_3
D_{2d}	Three C_2 axes, two σ_d planes, and one S_4 (coincident with one C_2)	$H_2C{=}C{=}CH_2$
T_d	Three C_2 axes ⊥ to each other, four C_3, six σ, and three S_4 containing C_2	CH_4, $SiCl_4$

[a]All point groups possess the identity element, *E*

Scheme 1.3 Flowchart for assigning the group symmetry ([1], p. 10)

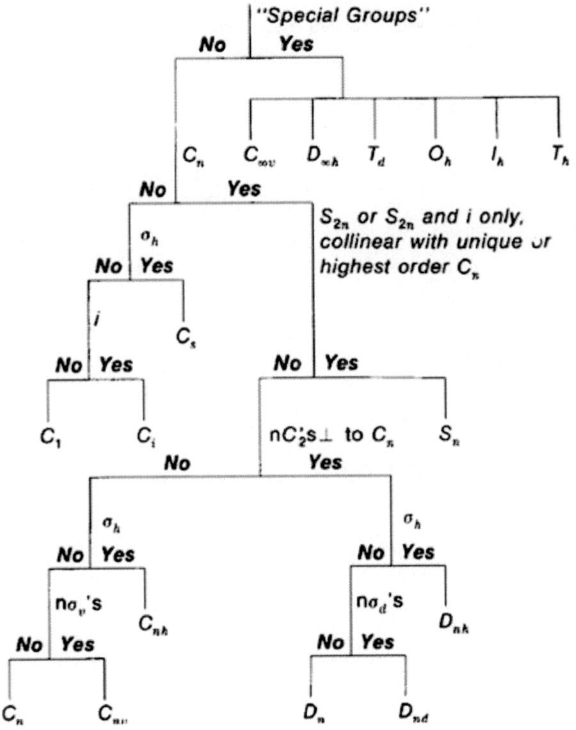

The elements in one group undergo rules as if they were numbers. The following paragraphs describe the properties of the group elements.

1.5.1 Rules for the Elements Which Constitute a Group

1. The product of groups elements must still produce a group element.
   For product, it is intended the combination of two operations, performed in a given order.
   e.g., C_2 x σ_v or σ_v x C_2
2. It is mandatory the existence of one operation commuting with the others and leaving them unchanged. This is called identity element

$$E \times \sigma_2 = \sigma_2 \times E = \sigma_2$$

3. The associative property must be applicable

$$AB(C) = A(BC)$$

4. Each element must have a reciprocal among the group elements. For each symmetry operation, there will be another one which anneals the first. C_3 has C_3^2 as reciprocal element; thus, C_3 x $C^2{}_3$ = 1.

1.5.2 Group Multiplication Tables

Let us consider the ammonia molecule NH_3 and its symmetry operations E, $2C_3$, $3\sigma_v$ in the C_{3v} group.

If we examine the combination of two symmetry operations, Table 1.5 matrix is obtained.

Table 1.5 Combination of symmetry operations in NH_3

	E	C_3	C_3^2	σ_y	σ_υ'	σ_υ''
E	E	C_3	C_3^2	σ_υ	σ_υ'	σ_υ''
C_3	C_3	C_3^2	E	σ_υ''	σ_υ	σ_υ'
C_3^2	C_3^2	E	C_3	σ_υ'	σ_υ''	σ_υ
σ_υ	σ_υ	σ_υ'	σ_υ''	E	C_3	C_3^2
σ_υ'	σ_υ'	σ_υ''	σ_υ	C_3^2	E	C_3
σ_υ''	σ_υ''	σ_υ	σ_υ'	C_3	C_3^2	E

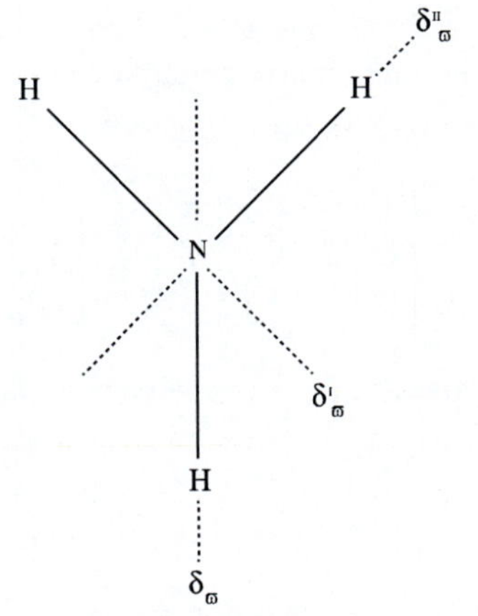

(σ_v, σ'_v, σ''_v indicate the symmetry planes).
Symmetry planes of NH_3

The matrix elements (Table 1.5) indicate that, following the rules given for the group elements, the group can be decomposed in subgroups which are smaller groups of elements satisfying all the requirements of a pristine group; e.g., the two subgroups are $E\,C\,C^2_{3v}$ $\sigma_v \sigma'_v \sigma''_v$.

1.5.3 Representation of Symmetry Operations by Matrices

The symmetry operations can be suitably represented by matrices.

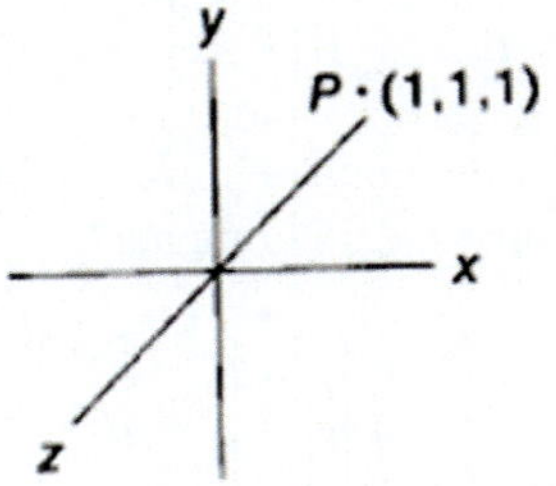

For the point P having the Cartesian $x\ y\ z$ coordinates, when it undergoes the identity operation, the following matrix notation can be assumed, where the ternary matrix corresponds to the identity operation and $X'\ Y'\ Z'$ represents

$$\begin{bmatrix} 1 & 0 & 0 \\ 0 & 1 & 0 \\ 0 & 0 & 1 \end{bmatrix} \begin{bmatrix} X \\ Y \\ Z \end{bmatrix} = \begin{bmatrix} X' \\ Y' \\ Z' \end{bmatrix}$$

the vector undergone the identity symmetry operation
that leads

$$X = X'\ Y = Y\ Z' = Z'$$

Analogously, the result of the operation σ_{yz} is

$$\begin{bmatrix} -1 & 0 & 0 \\ 0 & 1 & 0 \\ 0 & 0 & 1 \end{bmatrix} \begin{bmatrix} X \\ Y \\ Z \end{bmatrix} = \begin{bmatrix} X' \\ Y' \\ Z' \end{bmatrix}$$

That of σ_{xz} is

$$\begin{bmatrix} 1 & 0 & 0 \\ 0 & -1 & 0 \\ 0 & 0 & 1 \end{bmatrix} \begin{bmatrix} X \\ Y \\ Z \end{bmatrix} = \begin{bmatrix} X' \\ Y' \\ Z' \end{bmatrix}$$

For the inversion operation, the equation is

$$\begin{bmatrix} -1 & 0 & 0 \\ 0 & -1 & 0 \\ 0 & 0 & -1 \end{bmatrix} \begin{bmatrix} X \\ Y \\ Z \end{bmatrix} = \begin{bmatrix} X' \\ Y' \\ Z' \end{bmatrix}$$

For the clockwise rotation of the point x, y (vector P) by a φ angle, the rotation around the z-axis does not change the z component

$$\begin{bmatrix} & & 0 \\ & & 0 \\ 0 & 0 & 1 \end{bmatrix} \begin{bmatrix} x \\ y \\ z \end{bmatrix} = \begin{bmatrix} x' \\ y' \\ z' \end{bmatrix}$$

As for the x and y components, each of them becomes a linear combination of both x- and y-components.

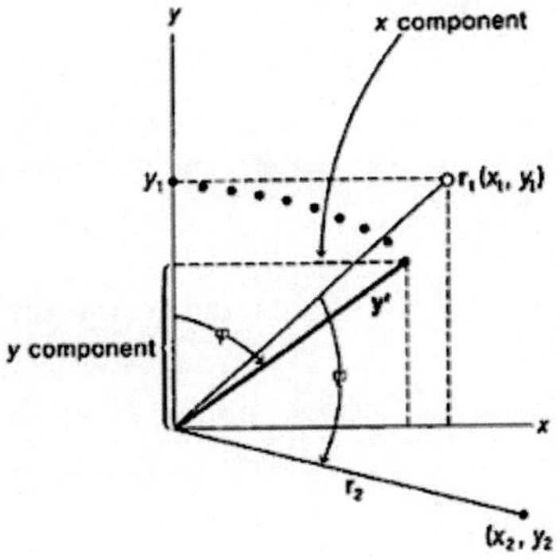

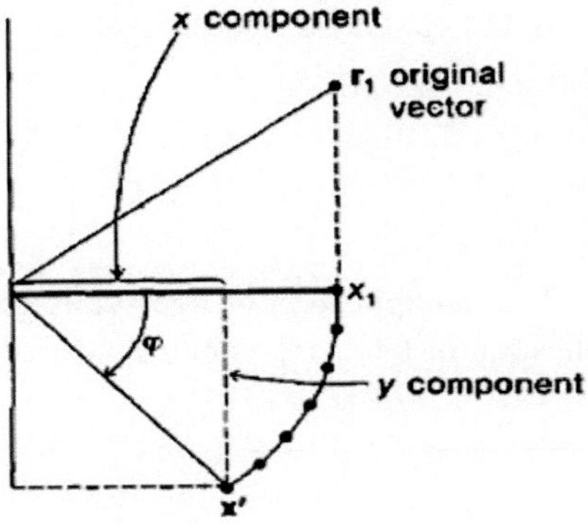

Rotation of the y-coordinate

Rotation of the x-coordinate

Rotation of the z-coordinate

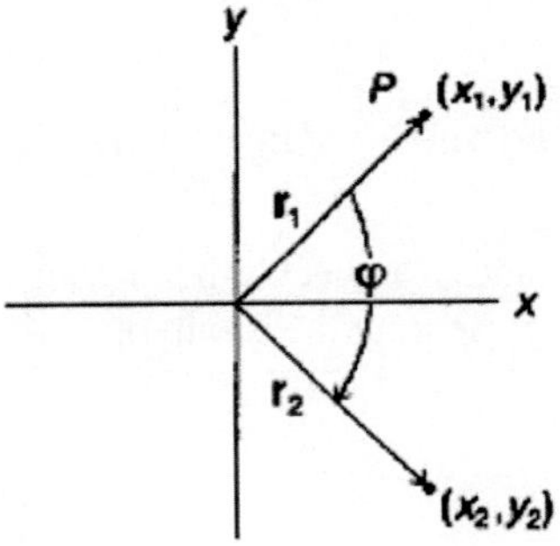

Point P in Cartesian coordinates

In total for both coordinates

$$x_2 = x_1 \cos \varphi + y_1 \sin \varphi$$
$$y_2 = -x_1 \sin \varphi + y_1 \cos \varphi$$

Writing in a matrix form these equations for a clockwise proper rotation of two coordinates,

$$\begin{bmatrix} \cos \varphi & \sin \varphi \\ -\sin \varphi & \cos \varphi \end{bmatrix} \begin{bmatrix} x_1 \\ y_1 \end{bmatrix} = \begin{bmatrix} x_2 \\ y_2 \end{bmatrix}$$

For all three coordinates,

$$\begin{bmatrix} \cos \varphi & \sin \varphi & 0 \\ -\sin \varphi & \cos \varphi & 0 \\ 0 & 0 & 1 \end{bmatrix}$$

In the particular case of C_2 rotation, the matrix becomes

$$\begin{bmatrix} -1 & 0 & 0 \\ 0 & -1 & 0 \\ 0 & 0 & 1 \end{bmatrix}$$

The multiplication of two symmetry operations can be associated to the multiplication of the corresponding matrices

$$\sigma_{xz} \times \sigma_{xz} = E$$

$$\begin{bmatrix} 1 & 0 & 0 \\ 0 & -1 & 0 \\ 0 & 0 & 1 \end{bmatrix} \begin{bmatrix} 1 & 0 & 0 \\ 0 & -1 & 0 \\ 0 & 0 & 1 \end{bmatrix} = \begin{bmatrix} 1 & 0 & 0 \\ 0 & 1 & 0 \\ 0 & 0 & 1 \end{bmatrix}$$

1.6 Irreducible Representations

Let us consider a molecule belonging to C_{2v} symmetry group (e.g., H_2O) and the matrices representing the group operations based on the three Cartesian coordinates

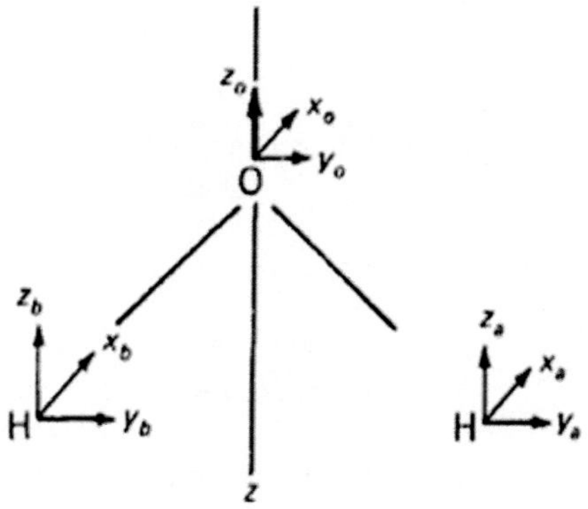

Local Cartesian coordinates in H_2O

$$E \quad C_2 \quad \sigma_{xz} \quad \sigma_{yz}$$

$$\overset{E}{\begin{bmatrix} 1 & & \\ & 1 & \\ & & 1 \end{bmatrix}} \overset{C_2}{\begin{bmatrix} -1 & & \\ & -1 & \\ & & 1 \end{bmatrix}} \overset{\sigma_{xz}}{\begin{bmatrix} 1 & & \\ & -1 & \\ & & 1 \end{bmatrix}} \overset{\sigma_{yz}}{\begin{bmatrix} -1 & & \\ & 1 & \\ & & 1 \end{bmatrix}}$$

Each matrix is composed of three one-dimensional matrices, which lie on the diagonal. If we are concerned with a one-dimensional vector like {x, 0, 0}, only the first row of the total set of the four matrices is necessary to represent the behavior of vector (1, −1, 1, −1). The row is called irreducible representation and takes the denomination of B_1. The irreducible representation for {0, y, 0} is (1, −1, −1, 1), labeled B_2 and that for {0, 0, z} is (1, 1, 1, 1) labeled A_1.

If a different object (function or operation) is selected to be base of the C_{2v} symmetry operations, e.g., the C_2 rotation, also the behavior of this object can be associated to an irreducible representation, in this case (1, 1, −1, −1), labeled A_2.

1.6.1 Character Table of groups

The so-called table of characters associated to the C_{2v} group is

	E	C_2	σ_{xz}	σ'_{yz}	
A_1	+1	+1	+1	+1	z
A_2	+1	+1	−1	−1	R_z
B_1	+1	−1	+1	−1	x, R_y
B_2	+1	−1	−1	+1	y, R_x

where the first column reports the name of the irreducible representation of the corresponding row and the last column indicates the base objects undergoing the symmetry transformations. As for the label of the irreducible representations:

A and B refer to representations which have one single function as base of the symmetry operations, A in particular has +1 character, and B has −1 for the C_n operation; 1 and 2 refer to the symmetric and antisymmetric role of the the σ_{xz} operation.

Let us consider the NH_3 molecule and its symmetry operations

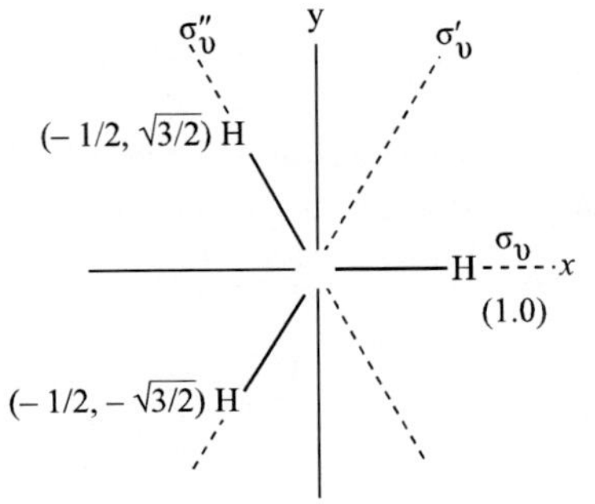

Symmetry planes of NH_3

$$E = \begin{bmatrix} 1 & 0 & 0 \\ 0 & 1 & 0 \\ 0 & 0 & 1 \end{bmatrix} \quad C_3 = \begin{bmatrix} -\frac{1}{2} & -\frac{\sqrt{3}}{2} & 0 \\ \frac{\sqrt{3}}{2} & -\frac{1}{2} & 0 \\ 0 & 0 & 1 \end{bmatrix}$$

For σ_v

$$\begin{bmatrix} 1 & 0 & 0 \\ 0 & -1 & 0 \\ 0 & 0 & 1 \end{bmatrix} \begin{bmatrix} X \\ Y \\ Z \end{bmatrix} = \begin{bmatrix} X' \\ Y' \\ Z' \end{bmatrix}$$

For σ'_v and σ''_v

$$\sigma_v' = \begin{bmatrix} -\frac{1}{2} & \frac{\sqrt{3}}{2} & 0 \\ \frac{\sqrt{3}}{2} & \frac{1}{2} & 0 \\ 0 & 0 & 1 \end{bmatrix} \quad \sigma_v'' = \begin{bmatrix} -\frac{1}{2} & -\frac{\sqrt{3}}{2} & 0 \\ -\frac{\sqrt{3}}{2} & \frac{1}{2} & 0 \\ 0 & 0 & 1 \end{bmatrix}$$

The table of characters for the C_{3v} group is

C_{3v}	E	$2C_3$	$3\sigma_v$		
A_1	1	1	1	z	x^2+y^2, z^2
A_2	1	1	−1	R_z	
E	2	−1	0	(x, y) $R_x R_y$	$(x^2-y^2, xy)(xz, yz)$
I		**II**		**III**	**IV**

In the first column is the representation label, where E is a bidimensional representation that means it has a double base function.

In the second, there are the trace of matrices, in the third and forth the base functions of the single representation.

1.6.2 *Factoring the Total Representations into Irreducible Representation*

Let us consider the H_2O molecule, having C_{2v} symmetry, and attribute 3 Cartesian coordinates to each atom. Looking at the symmetry representations, for the E operation, it results

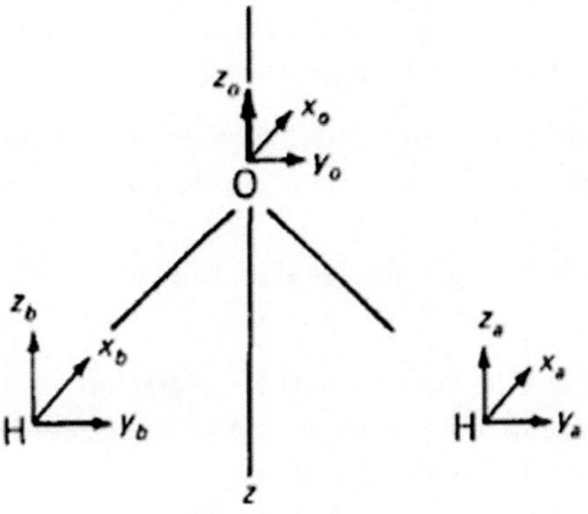

Local Cartesian coordinates in H_2O

$$\begin{bmatrix} 1 &&&&&&&& \\ & 1 &&&&&&& \\ && 1 &&&&&& \\ &&& 1 &&&&& \\ &&&& 1 &&&& \\ &&&&& 1 &&& \\ &&&&&& 1 && \\ &&&&&&& 1 & \\ &&&&&&&& 1 \end{bmatrix} \begin{bmatrix} x_o \\ y_o \\ z_o \\ x_a \\ y_a \\ z_a \\ x_b \\ y_b \\ z_b \end{bmatrix} = \begin{bmatrix} x'_o \\ y'_o \\ z'_o \\ x'_a \\ y'_a \\ z'_a \\ x'_b \\ y'_b \\ z'_b \end{bmatrix}$$

For the C_2 operation

$$\begin{bmatrix} -1 &&&&&&&& \\ & -1 &&&&&&& \\ && 1 &&&&&& \\ &&& 0 & 0 & 0 & -1 & 0 & 0 \\ &&& 0 & 0 & 0 & 0 & -1 & 0 \\ &&& 0 & 0 & 0 & 0 & 0 & 1 \\ &&& -1 & 0 & 0 & 0 & 0 & 0 \\ &&& 0 & -1 & 0 & 0 & 0 & 0 \\ &&& 0 & 0 & 1 & 0 & 0 & 0 \end{bmatrix} \begin{bmatrix} x_o \\ y_o \\ z_o \\ x_a \\ y_a \\ z_a \\ x_b \\ y_b \\ z_b \end{bmatrix} = \begin{bmatrix} x'_o \\ y'_o \\ z'_o \\ x'_a \\ y'_a \\ z'_a \\ x'_b \\ y'_b \\ z'_b \end{bmatrix}$$

similarly complicated matrices can be constructed for σ_{xz} and σ_{yz} operations. The traces of the matrices result

$$\begin{array}{lcccc} & E & C_2 & \sigma_{\upsilon_{xz}} & \sigma'_{\upsilon_{yz}} \\ \chi_T = & 9 & -1 & +1 & 3 \end{array}$$

They can be more easily obtained with the following rules, as alternative to construct the complete representation matrices:

1. Any vector which is unchanged by the symmetry operations contributes +1.
2. Any vector which changes into its opposite by the symmetry operations contributes −1.
3. Any vector which changes in another one under the symmetry operations contributes 0.

By these rules, it can be constructed the character of the total (reducible) representation. The character is used to decompose the reducible representation into irreducible ones by the following theorem.

$$a_i = \frac{1}{h}\sum_R g\chi_i(R)\chi_T(R)$$

where a_i is the number of irreducible representations of type i contained in the total one t, χ_i and χ_t are the characters under the different symmetry operations R added each other for all the operations, h is the number of the group operations, g the number of operations having the same character (class).

For the water molecule, having C_{2v} symmetry group

$$\begin{aligned} a_{A_1} &= \frac{1}{4}\left[g\chi_{A_i}(E)\chi_T(E) + g\chi_{A_i}(C_2)\chi_T(C_2) + g\chi_{A_i}(\sigma_v)\chi_T(\sigma_v) + g\chi_{A_i}(\sigma'_v)\chi_T(\sigma'_v)\right] \\ &= \frac{1}{4}[1\times 1\times 9 + 1\times 1\times(-1) + 1\times 1\times 1 + 1\times 1\times 3] = 3 \\ a_{A_2} &= \frac{1}{4}[1\times 1\times 9 + 1\times 1\times(-1) + 1\times(-1)\times 1 + 1\times(-1)\times 3] = 1 \\ a_{B_1} &= \frac{1}{4}[1\times 1\times 9 + 1\times(-1)\times(-1) + 1\times 1\times 1 + 1\times(-1)\times 3] = 2 \\ a_{B_2} &= \frac{1}{4}[1\times 1\times 9 + 1\times(-1)\times(-1) + 1\times(-1)\times 1 + 1\times 1\times 3] = 3 \end{aligned}$$

The total representation consists of the sum of $3A_1$ $1A_2$ $2B_1$ $3B_2$ irreducible representations, for a total number of nine representations, the same number as the initial coordinates in the molecule.

1.6.3 Relation Between Molecular Wave Functions and Irreducible Representations

Let us still consider that the energy of a molecule is unchanged by carrying out a symmetry operation, as well as the Hamiltonian too. Thus, a symmetry operator R commutes with the Hamiltonian operator.

Scheme 1.4 π orbitals in nitrite anion

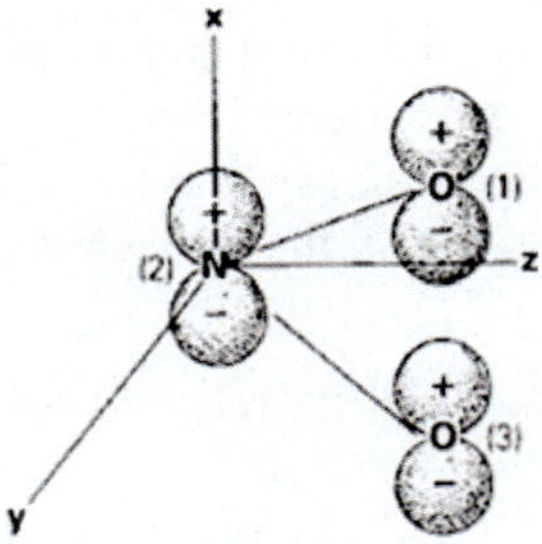

$$HR = RH$$

If we take the wave equation
$H\Psi_i = E\Psi_i$ and multiply each side by the symmetry operation R, we obtain
$HR\Psi_i = ER\Psi_i$ due to the commutation of operators
$R\Psi_i$ is the eigenfunction
$R\Psi_i = \pm 1\Psi_i$ and Ψ_i results basis of irreproducible representation.

1.6.4 Obtaining Molecular Orbitals with a Given Symmetry

Given the symmetry properties of the molecular orbitals, it appears not mandatory but very convenient that the molecular eigenfunctions were chosen as symmetry-adapted linear combination of atomic orbitals. They supply the best operative bases to perform the MO calculation, avoiding the long and unuseful calculations necessary in the case the eigenfunctions are not bases of irreducible representations. Let us consider the nitrite anion (Scheme 1.4).

Taking as operative basis the three p orbitals, in the C_{2v} symmetry group, the following values are obtained

C_{2v}	E	C_2	$\sigma_v(xz)$	$\sigma_v'(yz)$
	3	−1	+1	−3

to describe the behavior of the reducible representation. This later can be decomposed into $A_2 + 2B_1$ irreducible representations. We can now use the projection operators, working on the basis set of the three p orbitals, and annihilate any element in the basis set that does not contribute to a given irreducible representation. We generally work on an atomic orbital φ_i (i = 1, 2, 3), e.g., the p orbital of the an atom i, with the operator

$$\widehat{P} = \frac{l}{h}\sum_R \chi_R \widehat{R}$$

$\widehat{R}$ is the symmetry operator, χ_R is the character of the symmetry operator, l is the dimension of the irreducible representation, h is the group order (number of group operations). The sum is extended over all the symmetry operations. If we operate on φ_1 with the A_2 symmetry operator.

$$\psi(\text{for } A_2) = \widehat{P}(A_2)\varphi_1 = \frac{1}{4}[(1)E\varphi_1 + (1)C_2\varphi_1 + (-1)\sigma_v\varphi_1 + (-1)\,\sigma_v{}'\,\varphi_1]$$

The numbers in parenthesis are the elements of the A_2 irreducible representation, and φ_1 is the atomic orbital to be modified under the symmetry operation.

$$\begin{aligned}\widehat{P}(A_2)\varphi_1 &= \frac{1}{4}[(1)(1)\varphi_1 + (1)(-1)\varphi_3 + (-1)(1)\varphi_3 + (-1)(-1)\varphi_1] \\ &= N[\varphi_1 - \varphi_3]\end{aligned}$$

If we operate on the φ_2 orbital

$$\widehat{P}(A_2)\varphi_2 = \frac{1}{4}[(1)(1)\varphi_2 + (1)(-1)\varphi_2 + (-1)(1)\varphi_2 + (-1)(-1)\varphi_2] = 0$$

That means to remove (annihilate) the φ_2 orbital from the linear combination having A_2 symmetry.

The coefficient determination results.

$$\int (C_1\varphi_1 - C_1\varphi_3)^2 d\tau = C_1{}^2 \int \varphi_1{}^2 d\tau - 2\,C_1{}^2 \int \varphi_1\varphi_3 d\tau + C_1{}^2 \int \varphi_3{}^2 d\tau = 1$$
$$\int \varphi_1{}^2 d\tau = 1; \int \varphi_1\varphi_3 d\tau = 0; \quad \text{and} \quad \int \varphi_3{}^2 d\tau = 1.$$

$2C_1{}^2 = 1$ and the normalized function $(1/2)^{1/2}(\varphi_1 - \varphi_3)$.

No eigenfunctions with A_1 symmetry are expected from the decomposition of the reducible representation.

$$\widehat{P}(A_1)\varphi_1 = \frac{1}{4}[(1)(1)\varphi_1 + (1)(-1)\varphi_3 + (1)(1)\varphi_3 + (1)(-1)\varphi_1] = 0$$

Nor with B_2

$$\widehat{P}(B_2)\varphi_1 = \frac{1}{4}[(1)\varphi_1 + (+1)\varphi_3 + (-1)\varphi_3 + (-1)\varphi_1] = 0$$

Instead, two eigenfunctions with B_1 symmetry are expected

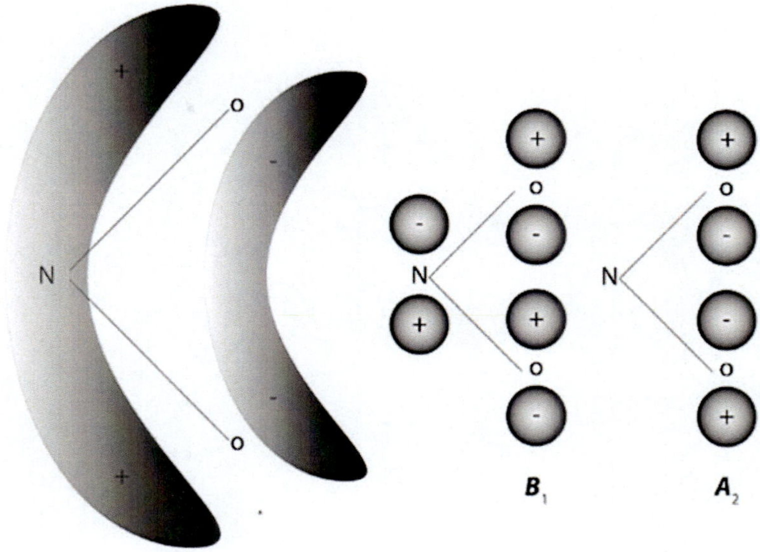

Scheme 1.5 Symmetry-adapted orbitals of nitrite anion

$$\widehat{P}(B_1)\varphi_1 = \frac{1}{4}[(1)(1)\varphi_1 + (-1)(-1)\varphi_3 + (1)(1)\varphi_3 + (-1)(-1)\varphi_1]$$
$$= N[\varphi_1 + \varphi_3]$$
$$\widehat{P}(B_1)\varphi_2 = \frac{1}{4}[(1)(1)\varphi_2 + (-1)(-1)\varphi_2 + (1)(1)\varphi_2 + (-1)(-1)\varphi_2] = \varphi_2$$

These two eigenfunctions with identical symmetry can mix as

$$\psi_1 = a\varphi_1 + b\varphi_2 + a\varphi_3$$
$$\psi_2 = a'\varphi_1 - b'\varphi_2 + a'\varphi_3$$

Significantly, the magnitude of a and b cannot be determined by only symmetry considerations; we can picture, however, the three symmetry-adapted molecular orbitals in terms of B_1 *ligand* orbital involving the in phase linear combination of three atomic p orbitals, B_1 *antiligand* orbital involving the out of phase linear combination, A_2 *non-ligand* orbital involving the combination of only two p atomic orbitals located on non-adjacent atoms. This can give a precise idea of the electronic distribution among the symmetry eigenfunctions built by π atomic orbitals (Scheme 1.5).

Chapter 2
Symmetry Orbitals for Different Bis and Polyatomic Molecules

The description will involve the examples of H_2 and the linear diatomic homonuclear molecules of the periodic table first row, of H_2O, NH_3, CO, CH_4 molecules. This is to demonstrate the power of group theory in predicting the chemical bond structure.

H_2 The molecule belongs to the $D_{\infty h}$ symmetry group, as well as all the diatomic homonuclear molecules of the periodic table first row, reported below. Assuming as atomic base function that constituted by two s orbitals, φ_1 and φ_2, the character tables for this base become

$D_{\infty h}$	E	C_{∞}^{φ}	σ_v	i	S_{∞}^{φ}	C_{2v}
χ_t	2	2	2	0	0	0

χ_t indicates the total representation of the base of the two s orbitals.

If the characters of the two irreducible representations Σ_g^+ and Σ_u^+ are considered, the total representation results in the sum of both $\chi_t = \Sigma_g^+ + \Sigma_u^+$

By using the projection operator, new symmetry eigenfunctions can be obtained

$D_{\infty h}$	E	C_{∞}^{φ}	σ_v	i	S_{∞}^{φ}	C_{2v}
$\sum_g^+$	1	1	1	1	1	1
$\sum_u^+$	1	1	1	−1	−1	−1
Φ_1	φ_1	φ_1	φ_1	φ_2	φ_2	φ_2
Φ_2	φ_2	φ_2	φ_2	φ_1	φ_1	φ_1

F. Morazzoni, *Symmetry in Inorganic and Coordination Compounds*, Lecture Notes in Chemistry 106, https://doi.org/10.1007/978-3-030-62996-0_2

$\sum_g^+ = \varphi_1 + \varphi_2$ (bonding) and $\sum_u^+ = \varphi_1 - \varphi_2$ (antibonding), in the following energy scheme

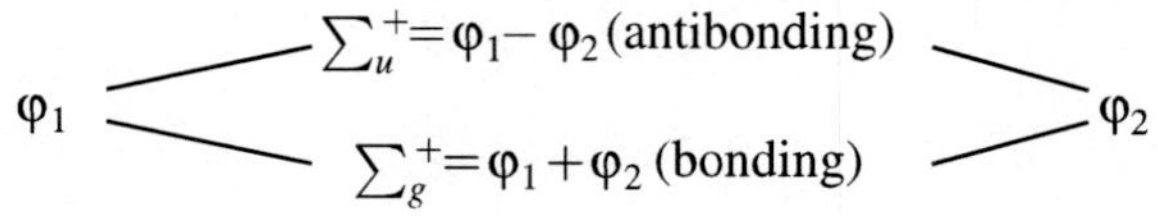

2.1 Homonuclear Diatomic Molecules A-A of the Periodic Table First Row

We assume the following base of atomic orbitals.

For the 2s orbitals

$\varphi_1 = 2s_A \quad \varphi_2 = 2s_B$ leads to

$\sum_g^+ = \varphi_1 + \varphi_2$ (bonding) $\sum_u^+ = \varphi_1 - \varphi_2$ (antibonding)

For the 2p orbitals $\varphi_1^A = 2p_x\ \varphi_2^A = 2p_y\ \varphi_3^A = 2p_z;\ \varphi_4^B = 2p_x\ \varphi_5^B = 2p_y\ \varphi_6^B = 2p_z$

$D_{\infty h}$	E	C_∞^φ	σ_v	i	S_∞^φ	C_{2v}
χ_t	6	$2+4\cos\varphi$	2	0	0	0

χ_t indicates the total representation of the base containing the six p orbitals.

Looking at the character tables of the linear groups $C_{\infty v}$ and $D_{\infty h}$

$C_{\infty v}$	E	$2C_\infty^\varphi$	...	$\infty\sigma_v$		
$A_1 \equiv \Sigma^+$	1	1	...	1	z	$x^2+y^2; z^2$
$A_2 \equiv \Sigma^-$	1	1	...	−1	R_z	
$E_1 \equiv \Pi$	2	$2\cos\Phi$	...	0	$(x; y) : (R_x; R_y)$	$(xz; yz)$
$E_2 \equiv \Delta$	2	$2\cos 2\Phi$	...	0		$(x^2-y^2; xy)$
$E_3 \equiv \Phi$	2	$2\cos 3\Phi$	...	0		
...	...	...	...	...		

$D_{\infty h}$	E	$2C_{\infty}^{\Phi}$	…	$\infty\sigma_v$	i	$2S_{\infty}^{\Phi}$	…	∞C_2		
Σ_g^+	1	1	…	1	1	1	…	1		$x^2+y^2; z^2$
Σ_g^-	1	1	…	–1	1	1	…	–1	R_z	
Π_g	2	$2\cos\Phi$	…	0	2	$-2\cos\Phi$	…	0	$(R_x; R_y)$	$(xz; yz)$
Δ_g	2	$2\cos 2\Phi$	…	0	2	$2\cos 2\Phi$	…	0		$(x^2-y^2; xy)$
…	…	…	…	…	…	…	…	…		
Σ_u^+	1	1	…	1	–1	–1	…	–1	z	
Σ_u^-	1	1	…	–1	–1	–1	…	1		
Π_u	2	$2\cos\Phi$	…	0	–2	$2\cos\Phi$	…	0	(x, y)	
Δ_u	2	$2\cos 2\Phi$	…	0	–2	$-2\cos 2\Phi$	…	0		
…	…	…	…	…	…	…	…	…		

and considering the character table of the irreducible representations of the $D_{\infty h}$ group

$D_{\infty h}$	E	C_{∞}^{φ}	σ_v	i	S_{∞}^{φ}	C_{2v}
$\sum_g^+$	1	1	1	1	1	1
$\sum_u^+$	1	1	1	-1	-1	-1
Π_g	2	$2\cos\varphi$	0	2	$-2\cos\varphi$	0
Π_u	2	$2\cos\varphi$	0	-2	$2\cos\varphi$	0

the base of six p atomic orbitals allows to build the following symmetry orbitals reported in a general Scheme 2.1 with the s-based atomic orbitals where the number before the symmetry orbital indicates the progression of orbitals of analogous symmetry.

$$\phi_{3\sigma_g^+} = (2p_{zA} + 2p_{zB})$$
$$\phi_{3\sigma_g^+} = (2p_{zA} - 2p_{zB})$$
$$\phi_{1\pi_u} = (2p_{xA} - 2p_{xB}) : \left(2p_{yA} + 2p_{yB}\right)$$
$$\phi_{1\pi_u} = (2p_{xA} - 2p_{xB}) : \left(2p_{yA} - 2p_{yB}\right)$$

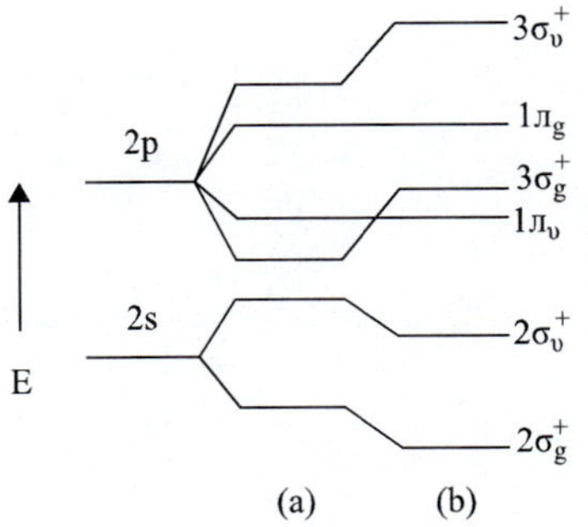

Scheme 2.1 General energy trend for the first row biatomic molecules

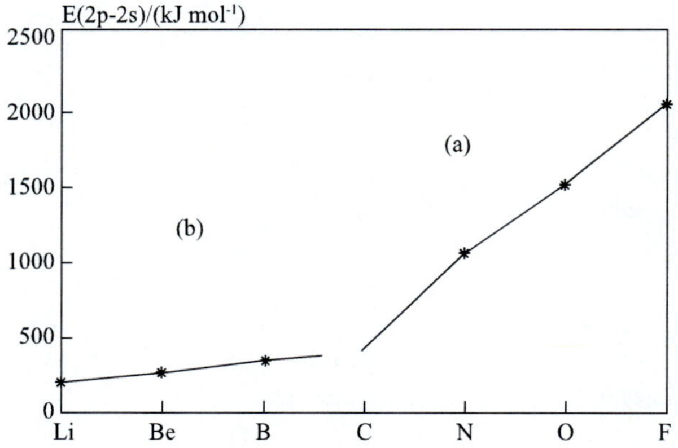

Scheme 2.2 Energy trend for the first row biatomic molecules as a function of the atomic species

The Scheme 2.2 shows the different energy trends for the first row biatomic molecules labeled **(a)** and for those labeled **(b)** and indicates that for the molecules **(b)** an energy inversion of σ_g^+ and π_u is active. This is connected to the small energy difference between s and p orbitals of the interacting atoms.

On proceeding along the first row of the periodic table, increasing the atomic number, the electronic population in Table 2.1 is expected, where the bond strength of the molecules is associated to the number and the type of populated symmetry orbitals.

Table 2.1 Orbital population for the first row biatomic molecules

Molecule/ion	$2\sigma_g^+$	$2\sigma_u^+$	$3\sigma_g^+$	$1\pi_u$	$1\pi_g$	$3\sigma_u^+$
Li_2	2					
Be_2	2	2				
B_2	2	2		2		
C_2	2	2		4		
N_2	2	2	2	4		
N_2^+	2	2	1	4		
N_2^-	2	2	2	4	1	
O_2	2	2	2	4	2	
O_2^+	2	2	2	4	1	
O_2^-	2	2	2	4	3	
O_2^{2-}	2	2	2	4	4	
F_2	2	2	2	4	4	
Ne_2	2	2	2	4	4	2

The overlap detail for the different orbitals is represented as

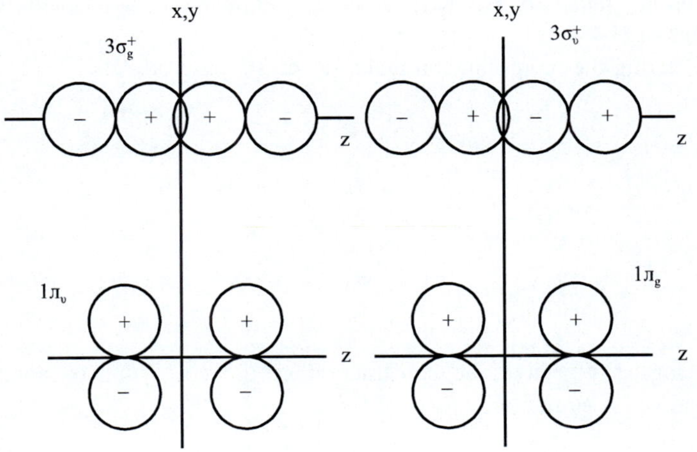

Overlap of p orbitals in omonuclear diatomic molecules

2.2 Water Molecule, as Example of Triatomic Molecule

We assume the following base of atomic orbitals $\varphi_1 = 1s_{H1}$ $\varphi_2 = 1s_{H2}$ $\varphi_3 = 1s_O$ $\varphi_5 = 2p_{Ox}$ $\varphi_6 = 2p_{Oy}$ $\varphi_7 = 2p_{Oz}$ (Scheme 2.3) in C_{2v} symmetry group

	E	C_2	σ(xz)	σ(yz)
χ_T	6	0	2	4
A_1	1	1	1	1
A_2	1	1	–1	–1
B_1	1	–1	1	–1
B_2	1	–1	–1	1

$\chi_T = 3\ A_1 + 2\ B_2 + B_1$

Scheme 2.3 Water molecule representation

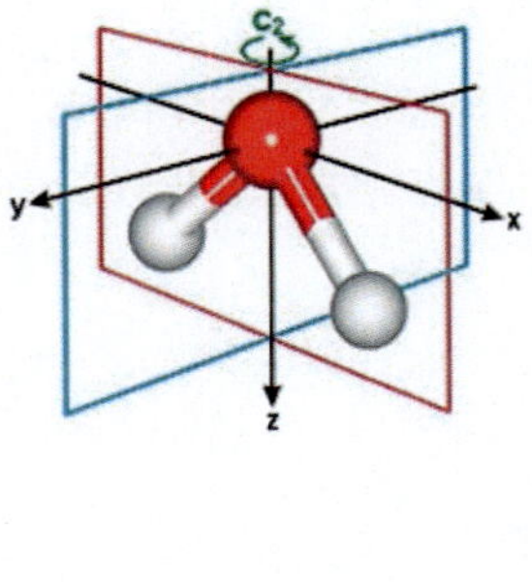

The φ_3 orbital has energy too different from the other ones to be conveniently combined by symmetry rules. The remaining ones give rise, by the use of the projection operator, to symmetry-adapted combinations associated to the C_{2v} symmetry of H_2O.

Considering the representation table for the six base orbitals

C_{2v}	E	C_2	$\sigma_v(xz)$	$\sigma_v(yz)$
φ_1	φ_1	φ_2	φ_2	φ_1
φ_2	φ_2	φ_1	φ_1	φ_2
φ_4	φ_4	φ_4	φ_4	φ_4
φ_5	φ_5	$-\varphi_5$	φ_5	$-\varphi_5$
φ_6	φ_6	$-\varphi_6$	$-\varphi_6$	φ_6
φ_7	φ_7	φ_4	φ_7	φ_7

and having as objective the construction of the six symmetry-adapted linear combinations of atomic orbitals (SALC)

$2A_1$ $\varphi_1 + \varphi_2 + \varphi_4$; $\varphi_7 \rightarrow \varphi_1 + \varphi_2 + \varphi_4$

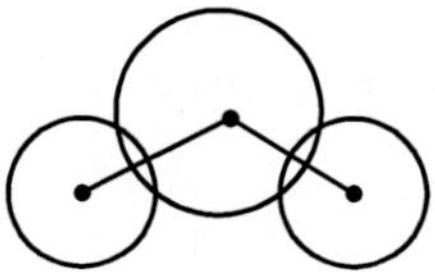

$1B_2$ $-\varphi_1 + \varphi_2 + \varphi_6$

$3A_1$ $\varphi_1 + \varphi_2 - \varphi_7 - \varphi_4$

$1B_1$ φ_5

$4A_1$ $\varphi_1 + \varphi_2 + \varphi_7 - \varphi_4$

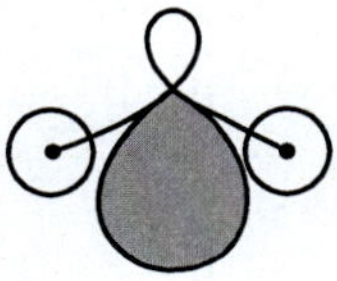

$2B_2$ $\varphi_1 - \varphi_2 + \varphi_6$

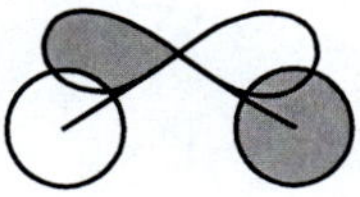

The numbers before the symmetry indicate the progression of orbitals of a given symmetry. The grey color indicates the negative sign of the function.

In the case we tentatively adopt a linear $D_{\infty h}$ symmetry (Scheme 2.4), the irreducible representation will be $2\sigma_g$, $1\sigma_u$, π_u, $3\sigma_g$, $2\sigma_u$, a lower number than for the C_{2v}, due to degeneracy of π_u. The lack of π_u induces stabilization of $3a_1$.

The HOH angle of 105.5° amplitude is the best convenient to minimize the energy of the water molecule. Scheme 2.5 reported the energy trend (A) and the H–O binding energy (B) against HOH angle.

The photoelectron spectrum of the water molecule (Fig. 2.1) confirms the symmetry energy trend, excluding the presence of degenerate π_u states.

Scheme 2.4 H_2O linear symmetry versus the bent one (the real one)

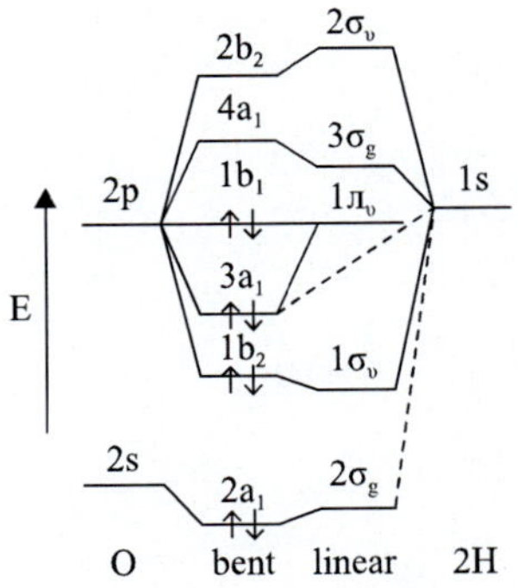

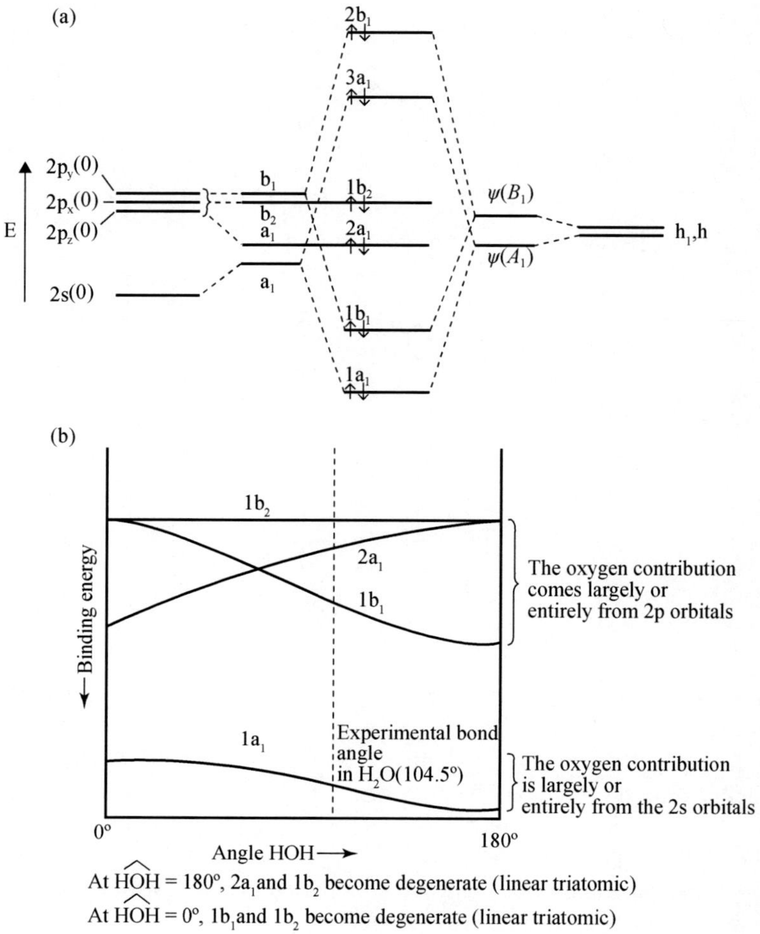

Scheme 2.5 **a** energy level trend of the water symmetry orbitals **b** the same against HOH angle amplitude. b_1 and b_2 are exchanged from (**a–b**)

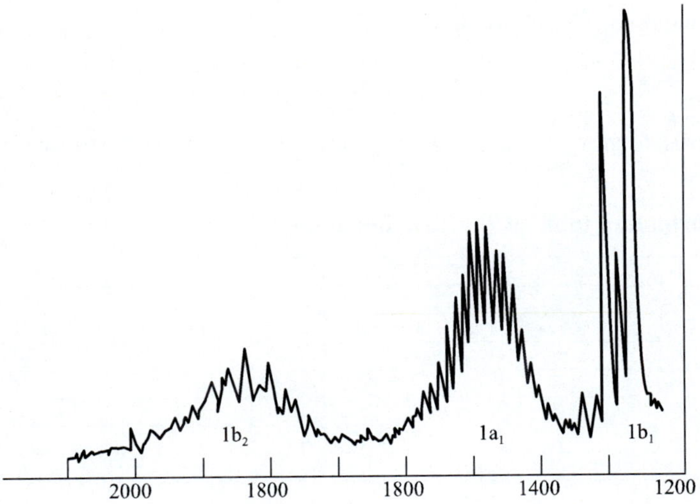

Fig. 2.1 Photoelectron spectrum of water, indicating the electronic transitions

Scheme 2.6 NH_3 molecule

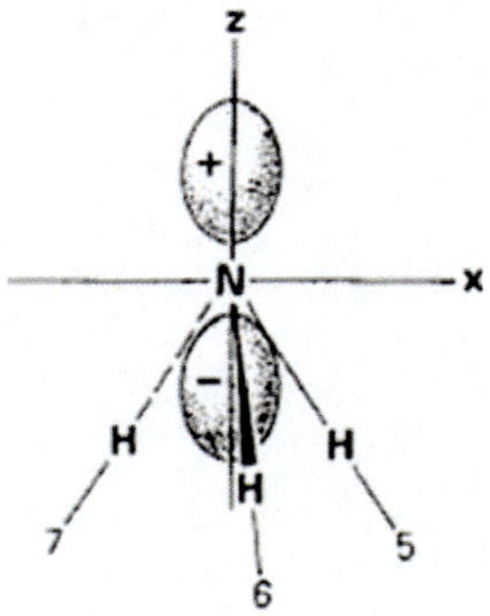

2.3 Ammonia Molecule, as Example of Tetratomic Molecule

We assume the following base of atomic orbitals $\varphi_1 = 2s_N$ $\varphi_2 = 2p_x$ $\varphi_3 = 2p_y$ $\varphi_4 = 2p_z$ $\varphi_5 = 1s\,(H_5)$ $\varphi_6 = 1s\,(H_6)$ $\varphi_7 = 1s\,(H_7)$ (Scheme 2.6).

Referring to C_{3v} symmetry, the characters of the reducible representation under the related symmetry operations are

E	$2C_3$	$3\sigma_v$
7	1	3

Decomposed into the irreducible ones:

$$3A_1 + 2E$$

In order to apply the projection operator, the following transformation table is compiled

Transformation table of the base functions

C_{3v}	E	C_3	C_3^2	$\sigma_v(H5)$	$\sigma_v(H6)$	$\sigma_v(H7)$
φ_1	φ_1	φ_1	φ_1	φ_1	φ_1	φ_1
φ_2	φ_2	$(-1/2\,\varphi_2 - \sqrt{3}/2\,\varphi_3)$	$(-1/2\,\varphi_2 + \sqrt{3}/2\,\varphi_3)$	φ_2	$(-1/2\,\varphi_2 + \sqrt{3}/2\,\varphi_3)$	$(-1/2\,\varphi_2 - \sqrt{3}/2\,\varphi_3)$
φ_3	φ_3	$(\sqrt{3}/2\,\varphi_2 - 1/2\,\varphi_3)$	$(-\sqrt{3}/2\,\varphi_2 - 1/2\,\varphi_3)$	$-\varphi_3$	$(\sqrt{3}/2\,\varphi_2 + 1/2\,\varphi_3)$	$(-\sqrt{3}/2\,\varphi_2 + 1/2\,\varphi_3)$
φ_4	φ_4	φ_4	φ_4	φ_4	φ_4	φ_4
φ_5	φ_5	φ_6	φ_7	φ_5	φ_7	φ_6
φ_6	φ_6	φ_7	φ_5	φ_7	φ_6	φ_5
φ_7	φ_7	φ_5	φ_6	φ_6	φ_5	φ_7

By using the projection operator, applied to either not degenerate or degenerate representations (A_1 and E), the following symmetry orbitals are obtained.

$$2A_1 = \varphi_1 + \varphi_5 + \varphi_6 + \varphi_7$$

$$1E_x = -\varphi_2 - 2\varphi_5 + \varphi_6 + \varphi_7$$
$$1E_y = \varphi_3 + \varphi_6 - \varphi_7$$

$1e_x$ $1e_y$

$$3A_1 = -\varphi_1 + \varphi_5 + \varphi_6 + \varphi_7 - \varphi_4$$

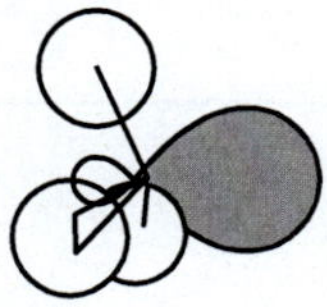

$$4A_1 = -\varphi_1 + \varphi_5 + \varphi_6 + \varphi_7 + \varphi_4$$

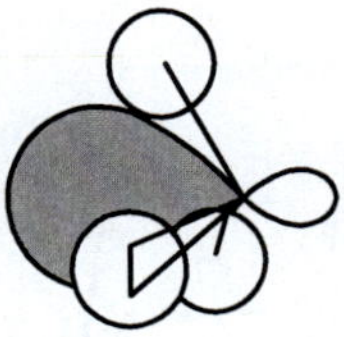

$$2E_x = 2\varphi_5 + \varphi_6 + \varphi_7 - \varphi_2$$
$$2E_y = \varphi_3 - \varphi_6 + \varphi_7$$

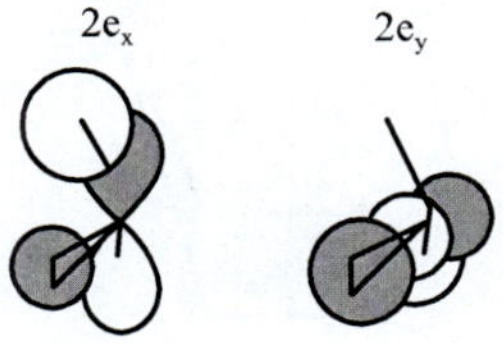

The Scheme 2.7 reports the energy level trends of ammonia for different possible symmetries. The stabilization of $3a_1$ energy level in C_{3v} symmetry, due to interaction with the ground state, is responsible for the stability of the pyramidal C_{3v} structure.

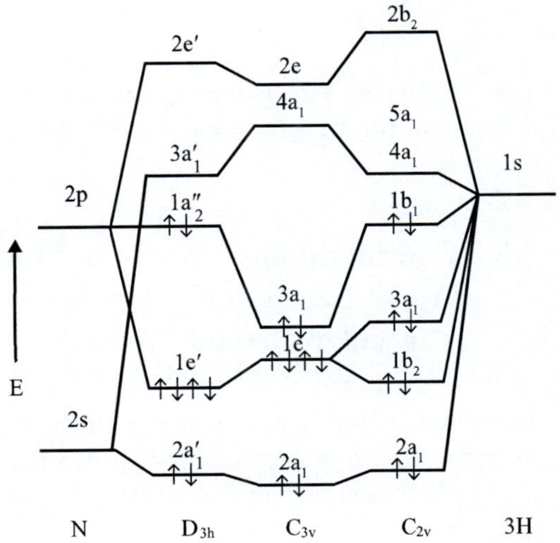

Scheme 2.7 Energy level trend of NH_3 symmetry orbitals

Orbital	Point group	
	T_d	D_{4h}
2s(C)	a_1	a_{1g}
$2p_x$(C)	t_2	e_u
$2p_y$(C)		
$2p_z$(C)		a_{2u}
4x1s(H)	$a_1 + t_2$	$a_{1g} + e_u + b_{1g}$

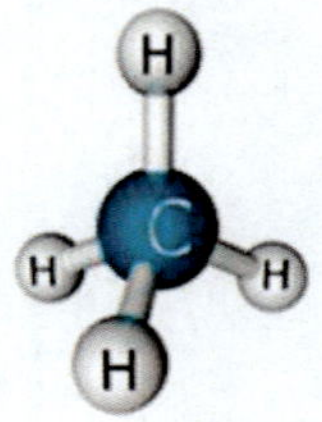

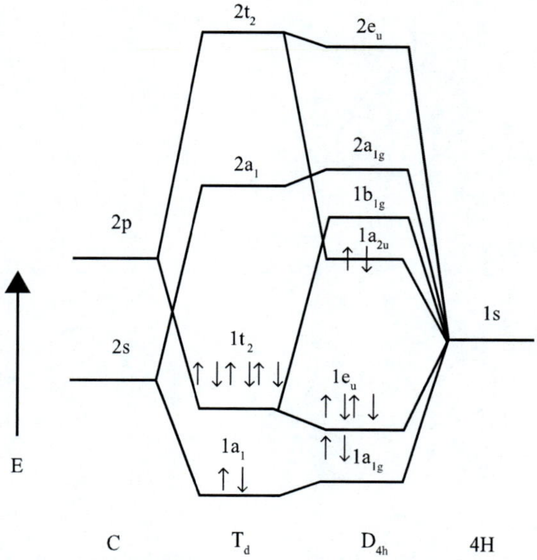

Scheme 2.8 Energy level trend of methane symmetry orbitals

2.4 Methane Molecule, as Example of Pentatomic Molecule

T_d symmetry guaranties the highest stability for the CH_4 unit due to the energy decrease of b_{1g} eigenfunction (Scheme 2.8).

2.5 Considerations About the Models of Chemical Bond in Use Before the Adoption of the Symmetry Molecular Orbital Model

In the time before the introduction of the molecular orbital model, two different models were proposed to rationalize the molecular shape, the reactivity, and the electronic configuration of a molecule. (1) the primitive Lewis model that does not consider any molecular shape and takes into account only the filling of the valence shell (8 electrons model); (2) the valence electron pair repulsion (VSEPR) theory which rationalizes the molecular shape in terms of the minimum pair repulsion, keeping the couples of bound atoms each ones separated in their interaction (also using hybrid orbitals).

The symmetry-adapted molecular orbitals takes as molecular orbitals those which are base of the irreproducible representation in the symmetry group of the molecule.

Step by step the model implementation goes from the only consideration of the electron number, to the total consideration of electrons and orbitals of the molecules, assessing that the bond is in charge to all the atomic orbitals, not only to those linking two atoms. By this way (see water molecule), both the shape and some reactivity can be rationalized. The spectroscopic experimental data (XPS) are in agreement with this symmetry approach.

It can be suggested that the symmetry constitutes a unifying approach to treat the chemical bond, as well as the molecular symmetry has the function of including all the constituents of the chemical bonds in the entire molecule. No energy is implied in the symmetry operations, and thus, the symmetry is only a way to describe in an easy way the electronic properties and to suggest that the symmetry is a condition to whom the electrons must obey.

As for the unique value of the symmetry, it has been invoqued also in the music composition; it was suggested that several composers, mainly Ludwing Van Beethoven, assembled the musical text by using measures related by symmetry operations. This guaranties the correlation among the various sections of the music score and the unit of it.

Chapter 3
Perturbation Theory

In this approximation, the hypothesis on Ψ_n is that these are built by known solutions Ψ^0 of the equation

$$H^0\Psi_n^0 = E_n^0\Psi_n^0$$

The hypothesis on H is that

$H = H^0 + H'$ where H' is the operator associated to the perturbation energy with respect to pristine H^0.

The following equation has to be solved

$$H\Psi_n = E_n\Psi_n$$

$$E_n = E_n^0 + \langle\Psi_n^0|H'|\Psi_n^0\rangle$$

$$\Psi_n = \Psi_n^0 + \sum_{k\neq n}\langle\Psi_k^0|H'|\Psi_n^0\rangle \times \Psi_k^0/E_n^0 - E_k^0$$

Evidently, the perturbed energy is the addition of the unperturbed one and of the perturbation energy obtained by the H' operator.

The perturbed eigenfunctions are the unperturbed ones modified by the perturbation energy, weighted on the energy differences between the ground state and the excited interacting states.

Assuming as example a system whose unperturbed eigenfunctions are Ψ_1^0 and Ψ_2^0, corresponding to energies E_1^0 and E_2^0

the determinant

$$\begin{vmatrix} H_{11} - E & H_{12} \\ H_{21} & H_{22} - E \end{vmatrix}$$

F. Morazzoni, *Symmetry in Inorganic and Coordination Compounds*, Lecture Notes in Chemistry 106, https://doi.org/10.1007/978-3-030-62996-0_3

where $H_{11} = \langle \Psi_1^0 | H | \Psi_1^0 \rangle \quad H_{22} = \langle \Psi_2^0 | H | \Psi_2^0 \rangle$

has H_{11} and H_{22} as solutions, when $H_{21} = H_{12}$,

$$
\begin{aligned}
E_1 &= H = \langle \Psi_1^0 | H^0 + H' | \Psi_1^0 \rangle = \langle \Psi_1^0 | H^0 | \Psi_1^0 \rangle + \langle \Psi_1^0 | H' | \Psi_1^0 \rangle \\
&= E_1^0 + \langle \Psi_1^0 | H' | \Psi_1^0 \rangle \\
E_2 &= H_{22} = E_2^0 + \langle \Psi_2^0 | H' | \Psi_2^0 \rangle
\end{aligned}
$$

In a similar way as in the case of LCAO method, it is still a great difficulty to exactly calculate the H_{ij} integrals. We will evaluate them by a semiempirical method, using the experimental parameters taken from the spectroscopic measurements.

The perturbation method can be positively applied when the perturbation energy is well lower than the energies of the unperturbed system, and it never suitably describes the covalent bond.

Based on this assumption, in the following, we will describe the perturbations that modellize the atomic electronic structure.

3.1 Interelectronic Repulsion Perturbation

If an atomic system has only one electron, there is no energy difference whatever is the orbital occupied by the electron. See for example one electron in the five d orbitals named by its m_l value

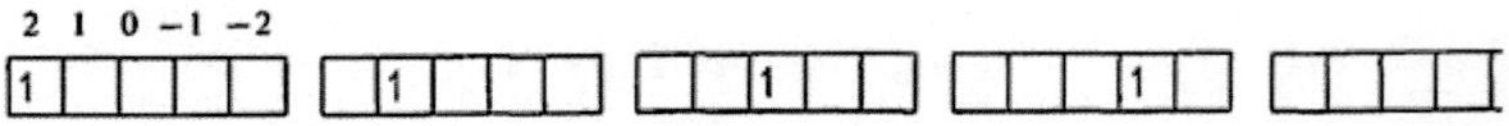

When the system contains two electrons, the energy level is depending on the relative occupation of the orbitals. It is necessary to combine all the possible electronic distributions by using the so-called vector atomic model. Table 3.1 reports all the possible combinations of two electrons into d orbitals (microstates), with their associated m_l value. In the columns, there is the sum of m_l values (M_L) related to the sum of l values (L). Also, m_s values of the individual electrons are reported as well as in the line there is the sum of the individual m_s (M_S).

In Table 3.1, the similar brackets include the terms of a given L value. The letters S, P, D, F, G, H, I corresponding to L = 0, 1, 2, 3, 4, 5, 6 indicate the effects of the electron–electron repulsion. Also, the spin multiplicity 2S + 1 is reported at the top left of the letter, and it results from the table.

In the case of d^1 electronic system

Table 3.1 Electronic microstates for the d^2 configuration

M_L \ M_S	(+ +) +1 (↑↑)	(+ −) 0 (↑↓)	(− −) −1 (↓↓)
4		2^+2^-	
3	(2^+1^+)	2^+1^- (2^-1^+)	(2^-1^-)
2	(2^+0^+)	2^+0^- (2^-0^+) 1^+1^-	(2^-0^-)
1	(2^+-1^+) $[1^+0^+]$	2^+-1^- (2^--1^+) 1^+0^- $[1^-0^+]$	$(2^--1^-)[1^-0^-]$
0	$(2^+-2^+)[1^+-1^+]$	2^+-2^- (2^--2^+) 1^+-1^- $[1^--1^+]$ 0^-0^+	$(2^--2^-)[1^--1^-]$

$$m_L = M_L = 2, 1, 0, -1, -2 \quad D \text{ term}$$
$$m_s = M_s = -1/2, +1/2 \quad S = 1/2 \text{ doublet term } {}^2D$$

In the case of d^2 electronic system

$$M_L = 4, 3, 2, 1, 0 \quad L = 4, 3, 2, 1, 0$$
$$L = 4 \quad M_L = -4, -3, -2, -1, 0, 1, 2, 3, 4 \quad G \text{ term}$$
$$L = 3 \quad M_k = -3, -2, -1, 0, 1, 2, 3 \quad F \text{ term}$$
$$L = 2 \quad M_k = -2, -1, 0, 1, 2 \quad D \text{ term}$$
$$L = 1 \quad M_k = -1, 0, 1 \quad P \text{ term}$$
$$L = 0 \quad M_k = 0 \quad S \text{ term}$$
$$S = 1 \quad M_s = -1, 0, +1 \quad {}^3F \; and \; {}^3P \text{ terms}$$
$$S = 0 \quad M_s = 0 \quad {}^1G \, {}^1S \, {}^1D \text{ terms}$$

Although the atomic vector model allows to predict the type of energy states (Scheme 3.1) that is the orbital and spin multiplicity, the scheme does not give the relative energies of the states. In order to calculate them, it is needed to apply the electronic repulsion operator.

How does it work?

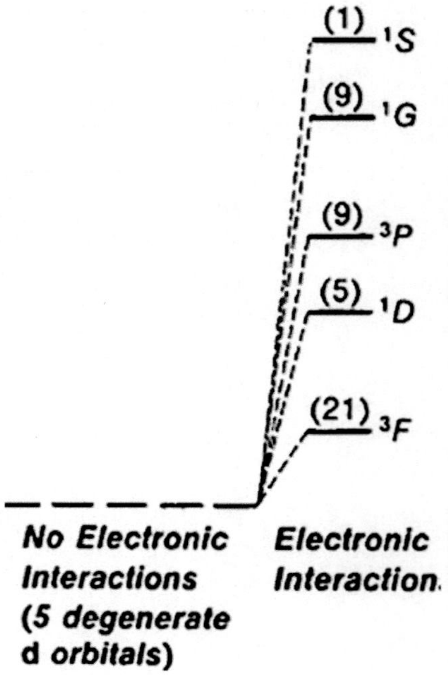

Scheme 3.1 Energy levels obtained for a d^2 configuration

The perturbation operator is

$$H' = \sum_{i<j} e^2/r_{ij}$$

The energies for all of the terms associated to the electronic configuration can be calculated and expressed by the Condon–Shortley parameters F_0 F_2 and F_4. These parameters are abbreviations for the various repulsion integrals in the perturbation determinant and allow the values can be determined by detecting the experimental atomic transitions. The energy expressions as a function of these parameters are independent of the atom, but the parameter values are there. The entire atomic spectrum is fitted by the same F parameter values.

For the V(III) ion d^2

$$E(^3P - {}^3F) = 15B$$
$$E(^1D - {}^3F) = 5B + 2C$$

$$B = 866\ \text{cm}^{-1} \text{ and } C/B = 3.6.$$

Racah redefined the Condon–Shortley parameters to simplify the energy separation between the states

Table 3.2 Free ion terms for various electronic configurations

n		Terms															
d^1	d^9	^2D															
d^2	d^8	^3F	^3P	^1G	^1D	^1S											
d^3	d^7	^4F	^4P	^2H	^2G	^2F	^2D	^2D	^2P								
d^4	d^6	^5D	^3H	^3G	^3F	^3F	^3D	^3P	^3P	^1I	^1G	^1G	^1F	^1D	^1D	^1S	^1S
d^5		^6S	^4G	^4F	^4D	^4P	^2I	^2H	^2G	^2G	^2F	^2F	^2D	^2D	^2D	^2P	^2S

$$B = F_2 - 5F_4 \quad C = 35F_4$$

Still for the V(III) ion

$$E(^3P - {}^3F) = 15F_2 - 75F_4 = 13{,}000\ \text{cm}^{-1}$$
$$E(^1D - {}^3F) = 5F_2 + 45F_4 = 10{,}600\ \text{cm}^{-1}$$

one obtains $F_2 = 1310\,\text{cm}^{-1}$ and $F_4 = 90\,\text{cm}^{-1}$.

The free ion terms for the different electronic configurations are in Table 3.2. Look at the analogy between the configurations complementary in 10. This originates from the analogy of the repulsion energy between two electrons and two holes. The magnitude of the electro–electron perturbation energy is few thousand cm^{-1}.

3.2 Spin–Orbit Coupling Perturbation

This perturbation describes the energy modification in an atom due to interaction of the angular spin magnetic moment with the angular orbital magnetic moment. Two schemes are used to treat the effect: the L.S scheme (Russel–Saunders) and the j-j scheme; the first one is the most common, and it is used when the spin–orbit perturbation is lower than the electronic repulsion (the major part of atoms); the second one is suitable to treat rare earth elements and the third-row transition elements.

The first case will be described in detail as it is the most common. The individual m_L values of the single electrons undergo coupling and produce the angular momentum L; the spin m_S values give the S value. The resultant momentum is called J, and it takes all the consecutive integer values spanning from L − S to L + S.

By still using the atomic vectorial model, it is easy to predict the type of spin–orbit-coupled states. In fact for the carbon atom

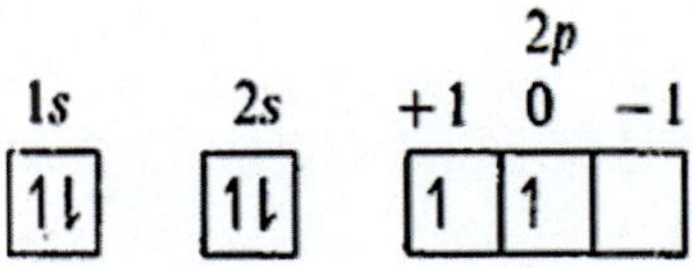

L = 1 S = 1 L − S = 0 L + S = 2 J = 0, 1, 2 the term is $^3P_0$ being the shell less than half filled.

Scheme 3.2 describes the spin–orbit coupling effects for d^2 configuration.

As for the spin–orbit coupling energy perturbation, two parameters, λ and ξ, are used to describe the energy magnitude.

$$\xi = \frac{Z_{\text{eff}} e^2}{2m^2c^2} \langle r^{-3} \rangle$$

where r^{-3} is the average value of the atomic radius, Z_{eff} is the nuclear charge, and m is the electron mass. The operator is

$$\xi L \cdot S$$

Using the parameter λ where

$$\lambda = \pm \xi / 2S$$

and if the shell is less than half filled, λ is positive, and the operator becomes λ L·S.

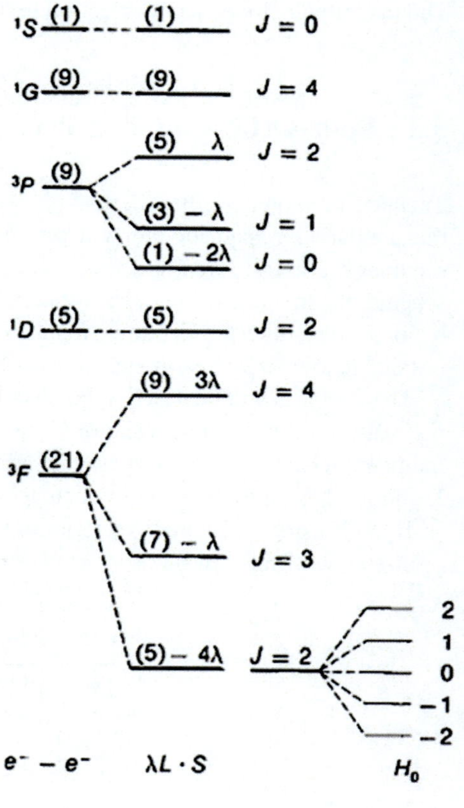

Scheme 3.2 Spin–orbit coupling effect in d^2 configuration

The spin–orbit terms, as in Scheme 3.2, are given by

$$\frac{1}{2}\lambda[J(J+1) - L(L+1) - S(S+1)]$$

The magnitude of the spin–orbit perturbation is generally few hundred cm^{-1}.

3.3 Crystal Field Perturbation

The perturbation described in this paragraph is able to model the bonding interaction between a central metal ion, with a given electronic configuration, and the surrounding groups (either ions or polar molecules). This way we shift from the atomic to a molecular context. Interestingly, the information on the bond is not related to the energy decrease of the metal electrons, but it involves the electronic repulsion between the electrons of the metal ion and the negative charge of the surrounding groups.

The base for the electronic perturbation in the case of transition metal ions is the following eigenfunctions

$$\begin{aligned} Y_2^0 &= (5/8)^{1/2}(3\cos^2\theta - 1)\cdot(2\pi)^{-1/2} \\ Y_2^{\pm1} &= (15/4)^{1/2}\sin\theta\cos\theta\cdot(2\pi)^{-1/2}e^{\pm i\varphi} \\ Y_2^{\pm2} &= (15/16)^{1/2}\sin^2\theta\cdot(2\pi)^{-1/2}e^{\pm 2i\varphi} \end{aligned}$$

Alternatively, in a real form, the more common called d orbitals are

$$\begin{aligned} d_{z^2} &= |0\rangle \qquad\qquad (d_{z^2} \text{ is really } d_{(z^2-r^2/3)} \\ d_{yz} &= (i/\sqrt{2})[|-1\rangle + |1\rangle] \\ d_{xz} &= (1/\sqrt{2})[|-1\rangle - |1\rangle] \\ d_{xy} &= -(i/\sqrt{2})[|2\rangle - |-2\rangle] \\ d_{(x^2-y^2)} &= (1/\sqrt{2})[|2\rangle + |-2\rangle] \end{aligned}$$

The complete Hamiltonian for the perturbed systems is

$$H = H_0 + V$$

where H_O is the free ion Hamiltonian and V the perturbation operator which describes the electronic repulsion between the ion electrons and the ligands, these simplified as point charges. For an octahedral interaction, the perturbation is

$$\widehat{V} = \sum_{i=1}^{6} eZ_i/r_{ij}$$

where e is the electron charge, Z_i the effective ligand charge, and r_{ij} the distance between electron and ligand. The form of the integrals in the perturbation determinant is $\langle M_L|V|M'_L\rangle$. When the integration is done, many quantities related to the radial part of the matrix elements appear in the determinant and have the form 1/6 $(Ze^2r^{-4}a^{-5})$ where r is the average radius of the d electrons of the central ion, and a the metal-ligand distance.

This quantity is referred to as *10 Dq* and is an energy. With these considerations, the determinant for an octahedral complex having a d^1 central ion is

$$\begin{array}{c} \\ |2\rangle \\ |1\rangle \\ |0\rangle \\ |-1\rangle \\ |-2\rangle \end{array} \begin{array}{ccccc} |2\rangle & |1\rangle & |0\rangle & |-1\rangle & |-2\rangle \\ \end{array}$$

$$\begin{array}{c|ccccc|} & |2\rangle & |1\rangle & |0\rangle & |-1\rangle & |-2\rangle \\ |2\rangle & Dq - E & & & & 5Dq \\ |1\rangle & & -4Dq - E & & & \\ |0\rangle & & & 6Dq - E & & \\ |-1\rangle & & & & -4Dq - E & \\ |-2\rangle & 5Dq & & & & Dq - E \end{array} = 0$$

This gives roots

$$E(|1\rangle) = -4Dq$$
$$E(|-1\rangle) = -4Dq$$
$$E(|0\rangle) = 6Dq$$

$$\begin{array}{c|cc|} & |2\rangle & |-2\rangle \\ |2\rangle & Dq - E & 5Dq \\ |-2\rangle & 5Dq & Dq - E \end{array} = 0$$

The residual determinant still has the roots $-4Dq + 6Dq$.

Thus, the crystal field effect separates the d orbitals in two sets, *10 Dq* energy distant, as in Scheme 3.3.

Scheme 3.3 Energy diagram of d^1 configuration in octahedral

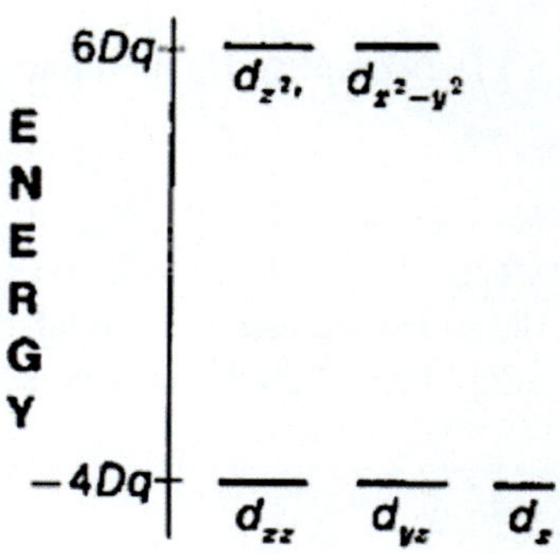

For the correlation between terms and d orbitals, see the beginning of this paragraph.

If we consider the d^2 configuration, many times discussed in the previous chapters, several free ion terms can be considered for the interaction with the octahedral crystal field. Let us start with the perturbation of the ^3F ground state. The secular determinant results

$$
\begin{array}{c|ccccccc|}
 & \langle 3| & \langle 2| & \langle 1| & \langle 0| & \langle -1| & \langle -2| & \langle -3| \\
\langle 3| & -3Dq-E & & & & 15^{1/2}Dq & & \\
\langle 2| & & 7Dq-E & & & & 5Dq & \\
\langle 1| & & & -Dq-E & & & & 15^{1/2}Dq \\
\langle 0| & & & & -6Dq-E & & & \\
\langle -1| & 15^{1/2}Dq & & & & -Dq-E & & \\
\langle -2| & & 5Dq & & & & 7Dq-E & \\
\langle -3| & & & 15^{1/2}Dq & & & & -3Dq-E
\end{array}
$$

These solutions are

$$
\begin{array}{ll}
E_1 = -6Dq & E_5 = 2Dq \\
E_2 = -6Dq & E_6 = 2Dq \\
E_3 = -6Dq & E_7 = 12Dq \\
E_4 = 2Dq &
\end{array}
$$

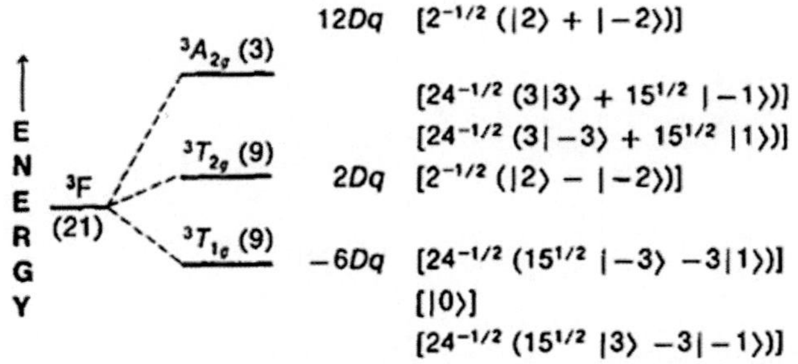

The denomination of the terms uses the group theory, in the way described in the next paragraph.

3.3.1 Use of the Group Theory to Predict the Orbital Spitting in a Given Symmetry

The target is how to be able assigning the terms obtained by the octahedral perturbation to the irreducible representations of the octahedral group itself. Given that the symmetry of the crystal field perturbation decides the form of the perturbation operator, it can be expected that the eigenfunctions of both singly or multi-electronic ions belong to the symmetry group of the complex. We begin by examining the effect of an octahedral field on the total representation whose d orbitals form the basis. Since all d orbitals have symmetry center, no new information arises from the symmetry inversion operation. Thus, it is easier to use the simple rotational group, called O group, which contains only the rotations present in the O_h group.

$$\begin{bmatrix} e^{2i\varphi} \\ e^{i\varphi} \\ e^{0} \\ e^{-i\varphi} \\ e^{-2i\varphi} \end{bmatrix} \underset{\text{by } \alpha}{\overset{\text{rotate}}{\longrightarrow}} \begin{bmatrix} e^{2i(\varphi+\alpha)} \\ e^{i(\varphi+\alpha)} \\ e^{0} \\ e^{-i(\varphi+\alpha)} \\ e^{-2i(\varphi+\alpha)} \end{bmatrix}$$

The vector contains as terms the five d orbitals, which undergo the rotation; notably only the φ angle varies, as the remaining polar coordinates, θ and ρ, are not changed by the symmetry operation.

The symmetry rotation is represented by the following matrix

$$\begin{bmatrix} e^{2i\alpha} & 0 & 0 & 0 & 0 \\ 0 & e^{i\alpha} & 0 & 0 & 0 \\ 0 & 0 & e^{0} & 0 & 0 \\ 0 & 0 & 0 & e^{-i\alpha} & 0 \\ 0 & 0 & 0 & 0 & e^{-2i\alpha} \end{bmatrix}$$

$$\begin{bmatrix} e^{li\alpha} & 0 & \cdots & 0 & 0 \\ 0 & e^{(l-1)i\alpha} & \cdots & 0 & 0 \\ \vdots & \vdots & \vdots & \vdots & \vdots \\ 0 & 0 & \cdots & e^{(1-l)i\alpha} & 0 \\ 0 & 0 & \cdots & 0 & e^{-li\alpha} \end{bmatrix}$$

more in general, whatever the base orbitals, the sum of the diagonal elements is

$$\chi(\alpha) = \frac{\sin\left(l + \frac{1}{2}\right)\alpha}{\sin\left(\frac{\alpha}{2}\right)}$$

In the case of a 120° rotation (C_3)

$$\chi(C_3) = \frac{\sin\left[\left(2 + \frac{1}{2}\right)\left(\frac{2\pi}{3}\right)\right]}{\sin\left(\frac{2\pi}{3\times 2}\right)} = \frac{\sin\frac{5\pi}{3}}{\sin\frac{\pi}{3}} = \frac{-\sin\frac{\pi}{3}}{\sin\frac{\pi}{3}} = -1$$

Using the above formula, we will represent the various operations in the O rotational group

	E	$6C_4$	$3C_2(=C_4^2)$	$8C_3$	$6C_2$
χ_T	5	−1	1	−1	1

X_T is decomposed in T_2 + E, in the case of O_h, T_{2g} + E_g.

The results obtained only by the symmetry considerations are the same as those deriving from the application of the perturbation operator, that is that the d orbitals split in two sets of orbitals, one three times degenerate (T_{2g}) and the other two times (E_g). The symmetry considerations allow to predict the number and type of energy levels, but do not give any quantitative energy measure. See the prevision in Table 3.3.

Let us consider that the correlation between the type of orbital (l value) and the irreducible representations is valid also for the states (L values) and the irreducible representations of the states.

The prevision of the states, both in the case of monoelectronic and multi-electronic states, can be transferred from the octahedral symmetry to the different ones. In fact, the same wave functions, in different point groups, are base of different irreducible representation (Table 3.4).

The crystal field perturbation can have different weight on the energy levels, that is it can be stronger (strong field) or weaker (weak field) than the interelectronic perturbation. If we suppose the d^n electronic configurations in a strong crystal field perturbation, for the d^1 configuration, the two perturbed terms are $^2T_{2g}$ and ^2E_g. In the case of d^2, we have t_{2g}^2, t_{2g}^1 e_g^1, e_g^2 orbital arrangements, and the terms are given by the direct product of the single arrangements (direct product of the irreducible representations).

$$
\begin{aligned}
t_{2g} \times t_{2g} &= T_{1g} + T_{2g} + E_g + A_{1g} \\
t_{2g}^1 \times e_g^1 &= T_{1g} + T_{2g} \\
e_g \times e_g &= E_g + A_{1g} + A_{2g}
\end{aligned}
$$

However, while the terms are easy to be obtained by the group theory, the related spin states are not. In fact, the d^2 configuration allows two spin states, the singlet and the triplet, and their attribution to the terms is not immediate. We will use the

Table 3.3 Association between orbitals and their symmetry properties in O_h

Types of orbital	*l* value	Irreducible representation
s	0	a_{1g}
p	1	t_{1u}
d	2	$e_g + l_{2g}$
f	3	$a_{2u} + t_{1u} + t_{2u}$
g	4	$a_{1g} + e_g + t_{1g} + t_{2g}$
h	5	$e_u + 2t_{1u} + t_{2u}$
i	6	$a_{1g} + a_{2g} + e_g + t_{1g} + 2t_{2g}$

Table 3.4 Correlation between the spectroscopic states in different symmetries

O_h	O	T_d	D_{4k}	D_{2d}	C_{4u}	C_{2u}	D_{3d}	D_3	C_{2h}
A_{1g}	A_1	A_1	A_{1g}	A_1	A_1	A_1	A_{1g}	A_1	A_g
A_{2g}	A_2	A_2	B_{1g}	B_1	B_1	A_2	A_{2g}	A_2	B_g
E_g	E	E	$A_{1g} + B_{1g}$	$A_1 + B_1$	$A_1 + B_1$	$A_1 + A_2$	E_g	E	$A_g + E$
T_{1g}	T_1	T_1	$A_{2g} + E_g$	$A_2 + E$	$A_2 + E$	$A_2 + B_1 + B_2$	$A_{2g} + E_g$	$A_2 + E$	$A_g + 2$
T_{2g}	T_2	T_2	$B_{2g} + E_g$	$B_2 + E$	$B_2 + E$	$A_1 + B_1 + B_2$	$A_{1g} + E_g$	$A_1 + E$	$2A_g +$
A_{1u}	A_1	A_2	A_{1u}	B_1	A_2	A_2	A_{1u}	A_1	A_u
A_{2u}	A_2	A_1	B_{1u}	A_1	B_2	A_1	A_{2u}	A_2	B_u
E_u	E	E	$A_{1u} + B_{1u}$	$A_1 + B_1$	$A_2 + B_2$	$A_1 + A_2$	E_u	E	$A_u + E$
T_{1u}	T_1	T_2	$A_{2u} + E_u$	$B_2 + E$	$A_1 + E$	$A_1 + B_1 + B_2$	$A_{2u} + E_u$	$A_2 + E$	$A_u + 2$
T_{2u}	T_2	T_1	$B_{1u} + E_u$	$A_2 + E$	$B_1 + E$	$A_2 + B_1 + B_2$	$A_{1u} + E_u$	$A_1 + E$	$2A_u +$

method of descending symmetry. If the e_g^2 configuration is considered, when we decrease the symmetry from O_h to D_{4h}, the determination of the spin state becomes straightforward, and the following configurations are expected

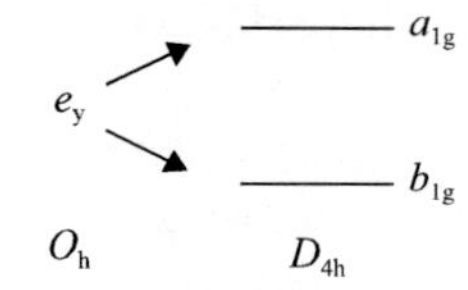

b_{1g}^2 (S = 0), a_{1g} b_{1g} (S = 0 and S = 1), a_{1g}^2 (S = 0) which give the terms A_{1g} (S = 0) B_{1g} (S = 0 and S = 1) A_{1g} (S = 0).

Given the following correlations:

O_h		D_{4h}
A_{1g}	→	A_{1g}
A_{2g}	→	B_{1g}
E_g	→	A_{1g}
	↘	B_{1g}

the spin state and the orbital degeneracy can be assigned

D_{4h}		O_h
${}^1A_{1g}$	→	$A_{1g}({}^1A_{1g})$
${}^3B_{1g}$	→	$A_{2g}({}^3A_{2g})$
${}^1A_{1g}$	→	$E_g({}^1E_g)$
${}^1B_{1g}$	↗	

The method of descending symmetry thus consists in rewriting the electronic configuration in a symmetry where all the orbital degeneracies disappear. By this way, the spin state can be definitely associated to the orbital. It results in a complete correlation diagram between the mono electronic configurations that correspond to very strong crystal field perturbation and the terms, where the electronic repulsion is stronger than the crystal field one (weak field).

Scheme 3.4 proposes the correlation between energies of weak and strong field and shows that these energies vary very much with the field strength, until to fully change the energy trend of the free ions states. However, there is a complete correspondence among high field and low spin terms.

It is important to observe that no crossover is possible, along the field variation, between the states which have the same symmetry and spin state, as the same energy of two states which derive from different configurations is nonsense.

Looking at the d^2 diagram, the configurations are associated to a number of electronic transitions which, under excitation energy, can change the electronic state of the ion and also allow to assign the ion to a given electronic configuration and field symmetry.

Based on the possibility of experimentally detecting the electronic transitions and consequently assigning the crystal field symmetry and the ion electronic configuration, taking also into consideration that the most probable (strong) transitions are those spin allowed, Orgel proposed the schemes (Scheme 3.5a, b).

The transition energies reported by these schemes, plotted for increasing *Dq* values, are between states with the same spin multiplicity and refer to octahedral

Scheme 3.4 Energy diagram for d^2 configuration in different field strengths

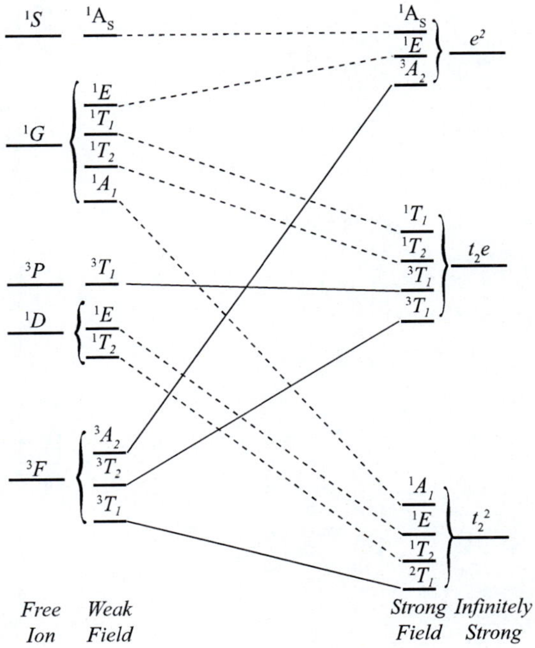

and tetrahedral symmetries. They allow to assign the crystal field of the spectra from d^1 to d^9 configurations and to recognize both symmetry and electronic configuration of the ion.

Tanabe and Sugano alternatively described the cubic (octahedral and tetrahedral) configurations by diagrams which report all the states for a single electronic configuration, independently of the similarity of their spin multiplicity. The diagrams plot the electronic energies of the ions states against the *Dq* energy values. All the energy values are measured in B (Racah) units in order to make the values independent of the different ions. On the left hand of the diagrams, we read the free ion energies, on the right hand the energy for an infinitely strong field. When the strong field configuration is not indicated, it is intended that it remains the same as the lower energy state (Scheme 3.6).

From the d^4 to d^7 configuration, the diagrams contain a vertical line which separates the high spin from the low spin region for the same ion. The two regions are different in the ground and the other states.

The transitions in Tanabe–Sugano diagrams can be read by the same way as for the Orgel diagrams and show the complete spectrum of the absorption bands.

Scheme 3.5 Orgel diagrams for the indicated electronic d configurations

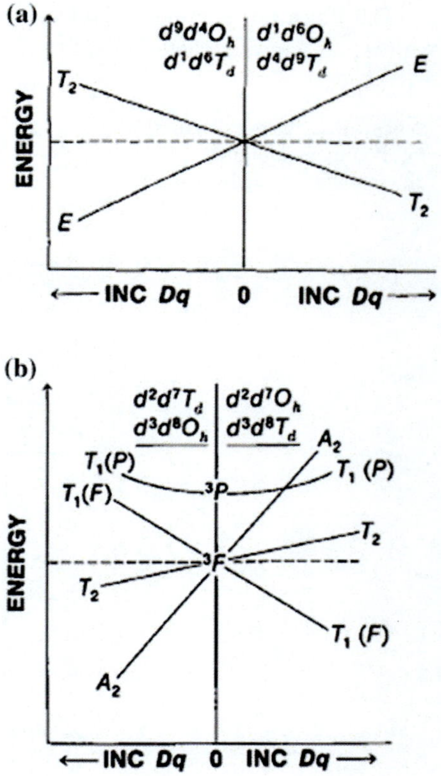

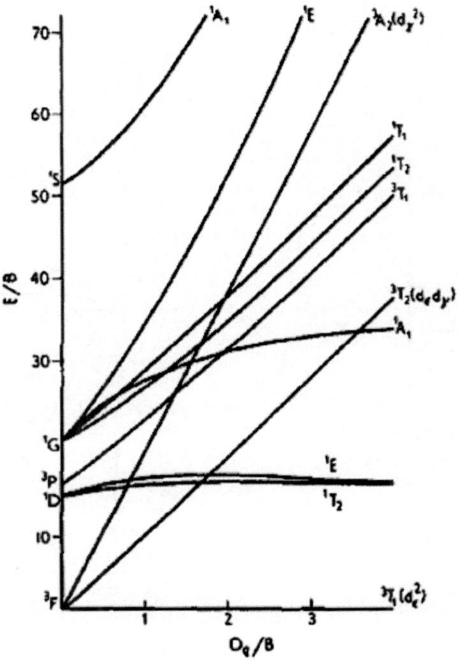

Energy diagram for the configuration d^2.

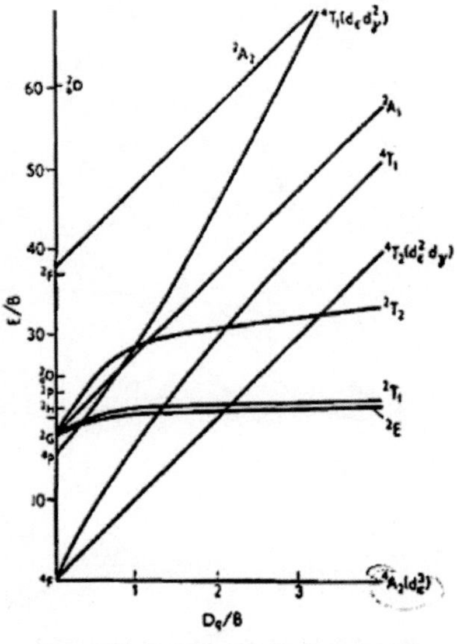

Energy diagram for the configuration d^3.

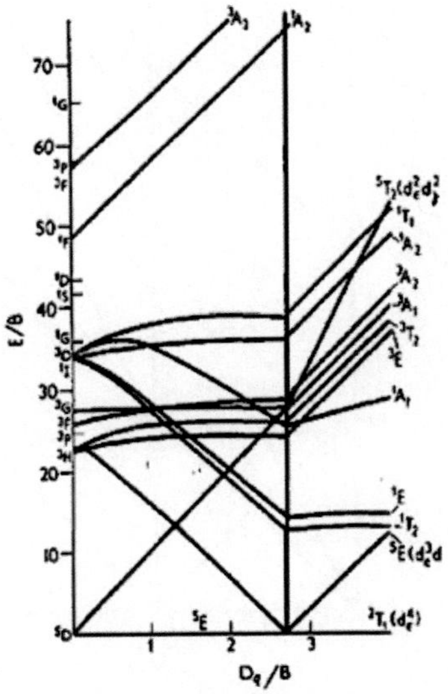

Energy diagram for the configuration d^4

Scheme 3.6 Tanabe–Sugano diagrams for different electronic configurations

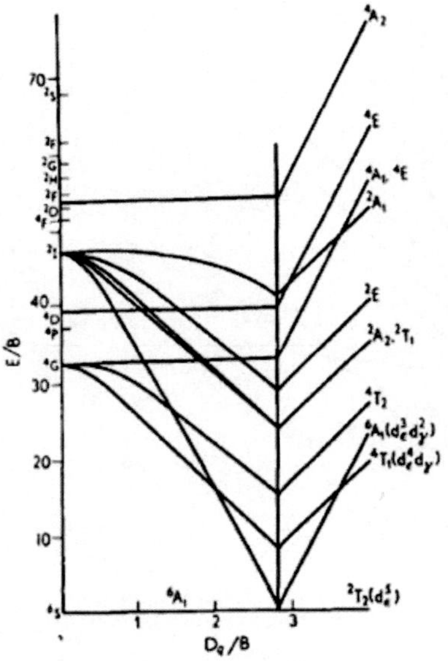

Energy diagram for the configuration d^5.

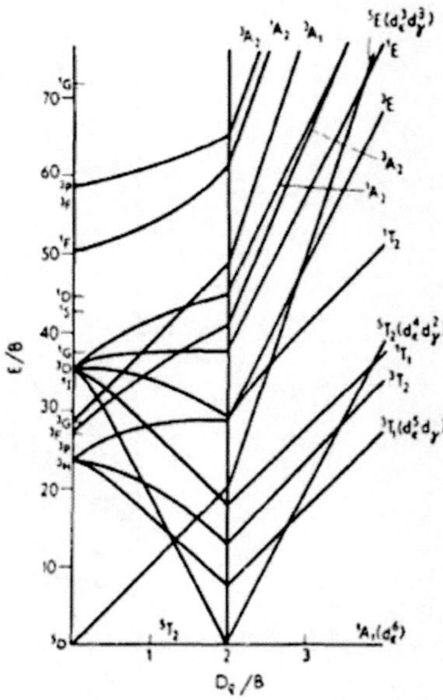

Energy diagram for the configuration d^6.

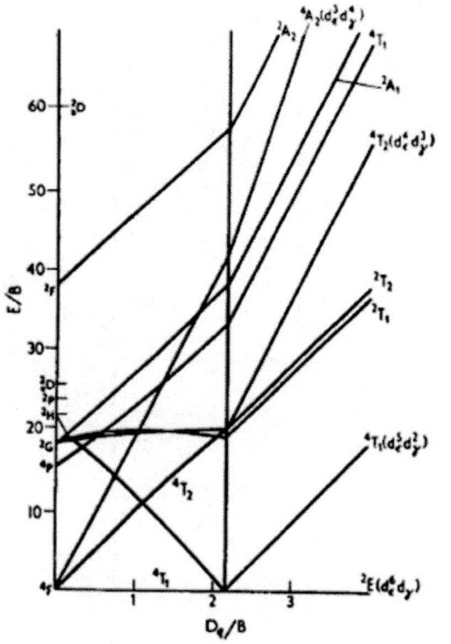

Energy diagram for the configuration d^7.

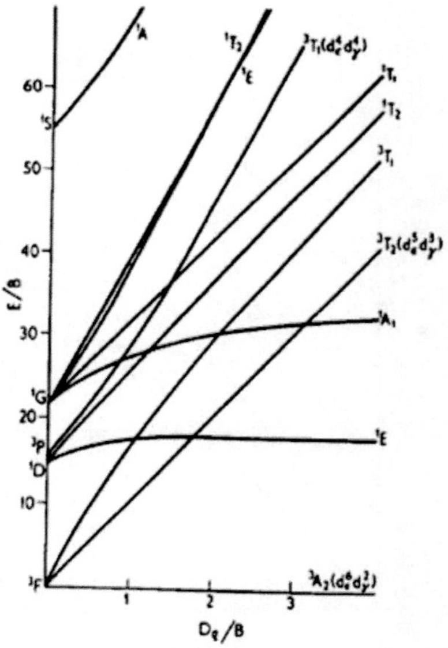

Energy diagram for the configuration d^8.

Scheme 3.6 (continued)

3.4 Examples of Attribution of the Absorption Bands

Let us consider the absorption maxima of some Ni(II) octahedral complexes (Fig. 3.1 and Table 3.5).

Three bands are evident, whose assignment is obtained by the Orgel diagram for the d^8 electronic configuration. The shapes of all the reported spectra are similar and indicate that the electronic configuration and the symmetry are similar. Only the energy of the transitions is different (see Table 3.5) and can be associated to the different crystal field strength.

As expected the presence of N in the first coordination Ni(II) sphere induces higher field strength, due to the stronger basicity of the ligand.

In Fig. 3.2, we report and assign the electronic transitions of a number of octahedral coordination compounds, by using the Tanabe–Sugano diagrams.

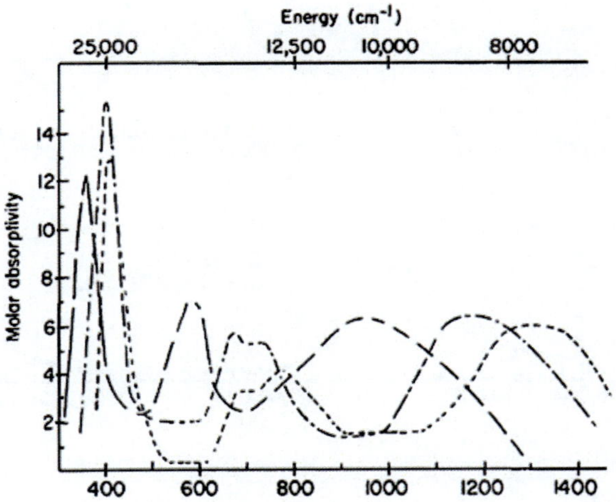

Fig. 3.1 Electronic absorption spectra of Ni(II) complexes ([1], p. 382)

Table 3.5 Crystal field transitions of Ni(II) octahedral complexes

Ligand	$^3A_{2g} \rightarrow {}^3T_{2g}$	$^3A_{2g} \rightarrow {}^3T_{1g}(F)$	$^3A_{2g} \rightarrow {}^3T_{1g}(P)$
H_2O	8500	15,400	26,000
NH_3	10,750	17,500	28,200
$(CH_3)_2SO$	7730	12,970	24,040
$HC(O)N(CH_3)_2$	8500	13,605 (14,900)	25,000
$CH_3C(O)N(CH_3)_2$	7575	12,740 (14,285)	23,810

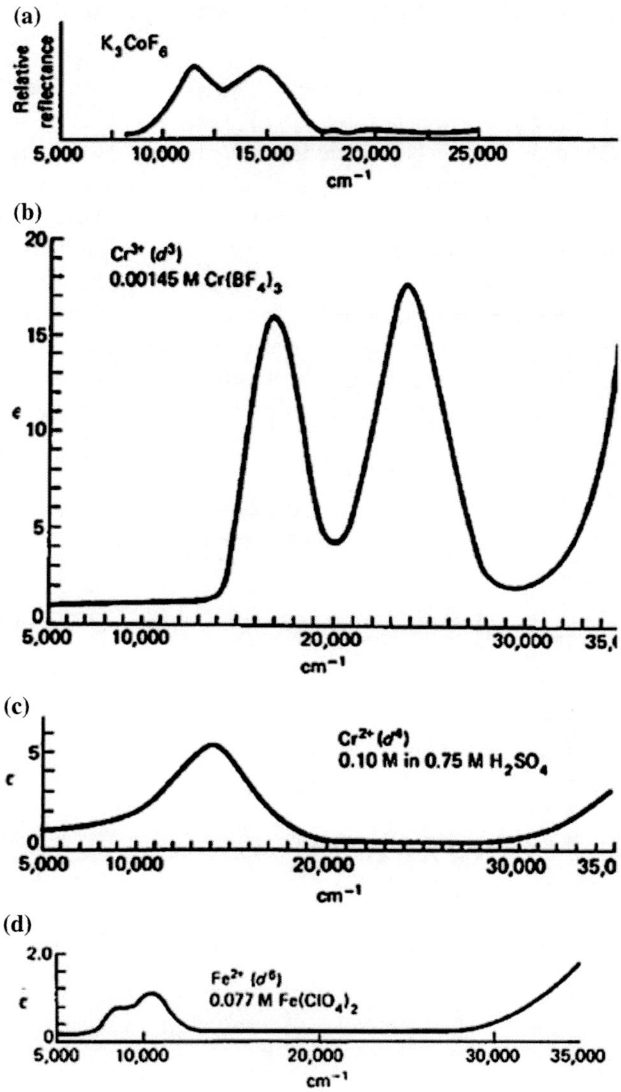

Fig. 3.2 Electronic spectra of different d^n configurations. The transitions are assigned basing on Tanabe–Sugano diagrams. **a** Co^{3+} d^6 $^5T_{2g} \rightarrow {}^5E_g$, **b** Cr^{3+} d^3 $^4A_{2g} \rightarrow {}^4T_{2g}$ $\;{}^4A_{2g} \rightarrow {}^4T_{1g}$ (F), **c** Cr^{2+} d^4 $^5E_g \rightarrow {}^5T_{2g}$, **d** Fe^{2+} d^6 $^5T_{2g} \rightarrow {}^5E_g$, **e** Mn^{2+} d^5 no spin allowed electronic transitions, **f** Ti^{3+} d^1 $^2T_{2g} \rightarrow {}^2E_g$, **g** V^{3+} d^2 $^3T_{1g}$ (F) $\rightarrow {}^3T_{2g}$ $\;{}^3T_{1g}$ (F) $\rightarrow {}^3T_{1g}$ (P), **h** Ni^{2+} d^8 $^3A_{2g} \rightarrow {}^3T_{2g}$ $\;{}^3A_{2g} \rightarrow {}^3T_{1g} \rightarrow$ (F) $\;{}^3A_{2g} \rightarrow {}^3T_{1g}$ (P), **i** Cu^{2+} d^9 $^2E_g \rightarrow {}^2T_{2g}$ ([1], pp. 406–408)

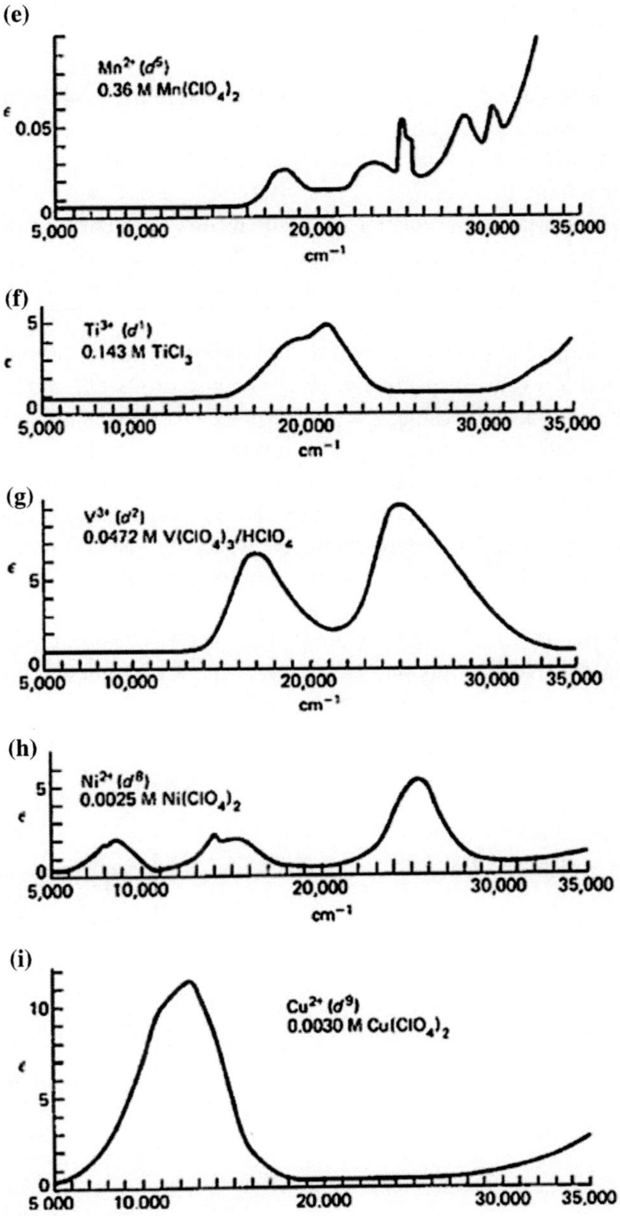

Fig. 3.2 (continued)

3.5 Crystal Field Limits

Several limitations affect the crystal field theory, due to the simplified model which considers the ligands as negative spheres, and the interaction between the valence electrons and the negative ligands charges a mere Coulomb repulsion. No covalent interaction is admitted.

For this reason, the experimental *Dq* value results different from those expected from the simple crystal field approximation. The diagrams show the variations of *Dq* when the overlap between σ or π ligand orbitals and d orbitals is taken into account.

Scheme 3.7 indicates the overlap of d orbitals with empty π orbitals (B) with the filled ones, and it results, respectively, in an increase of *Dq* (Δ) and a decrease with respect to the crystal field value.

Scheme 3.8 indicates the overlap of d orbitals with the σ ligand orbitals.

The covalent interaction leads to an increase of *Dq* value. Thus, it is possible to order the effects of the different ligands units based on the π and σ interactions they display. The trend is called spectrochemical series, and the comparison among the ligand effect has to be made only for the same metal ion.

$$I^- < Br^- < -SCN^- < F^- < \text{urea} < OH^- < CH_3CO_2^- < C_2O_4^{2-} <$$
$$H_2O < -NCS^- < \text{glycine} < C_5H_5N \sim NH_3 < \text{ethylenediamine} <$$
$$SO_3^{2-} < o\text{-phenanthroline} < NO_2^- < CN^-$$

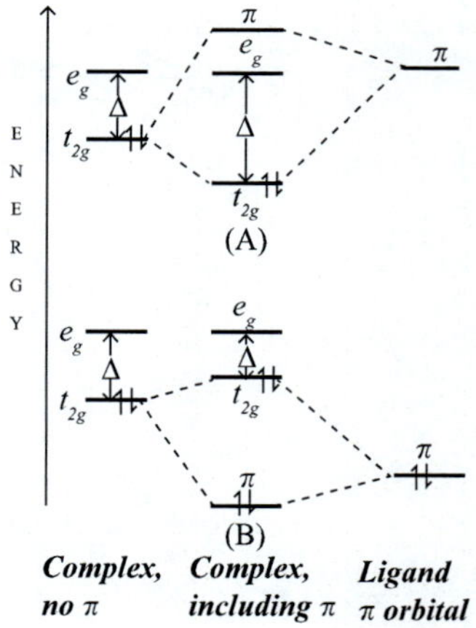

Scheme 3.7 Ligand field interaction of d orbitals with π ligand orbitals

Scheme 3.8 Ligand field interaction of d orbitals with σ ligand orbitals

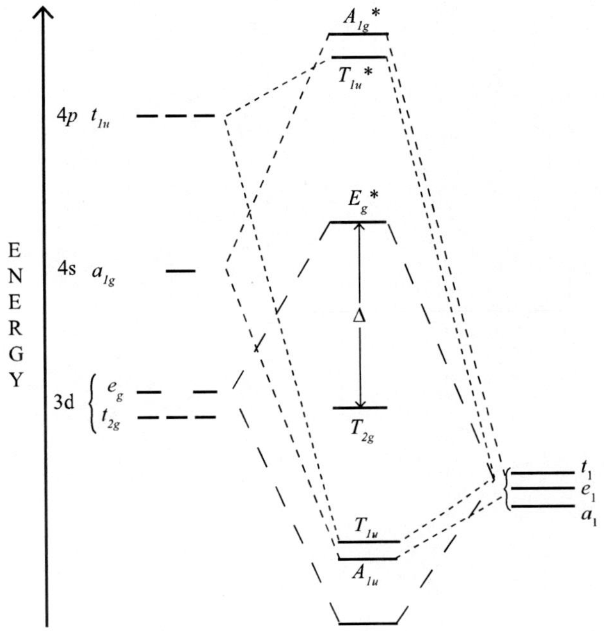

Instead the crystal field trend (based on the magnitude of 10 *Dq*) for different metal ions is

$$\mathrm{Mn(II)} < \mathrm{Ni(II)} < \mathrm{Co(II)} < \mathrm{Fe(II)} < \mathrm{V(II)} < \mathrm{Fe(III)} < \mathrm{Cr(III)} < \mathrm{V(III)} < \mathrm{Co(III)} < \mathrm{Mn(IV)} < \mathrm{Mo(III)} < \mathrm{Rh(III)} < \mathrm{Pd(IV)} < \mathrm{Ir(III)} < \mathrm{Re(IV)} < \mathrm{Pt(IV)}$$

By this way, we change the crystal field theory in ligand field theory.

3.6 Effect of Distortion from Cubic Symmetry

The group theory allows an easy indication of the crystal field effects on the atomic states, and we reported the examples in the previous chapters. However, the extensive treatment has been done for cubic symmetry. It is expected that in the case of lower symmetry, or distortion from the cubic one, the perturbation field effects on the ion states would be different. In fact for a d^1 ion in a distorted octahedral symmetry (elongated tetragonal), the pristine O_h states split as shown in the Scheme (upper part). For a low spin Co(III) d^6, the pristine O_h ground state changes as in the Scheme (lower part).

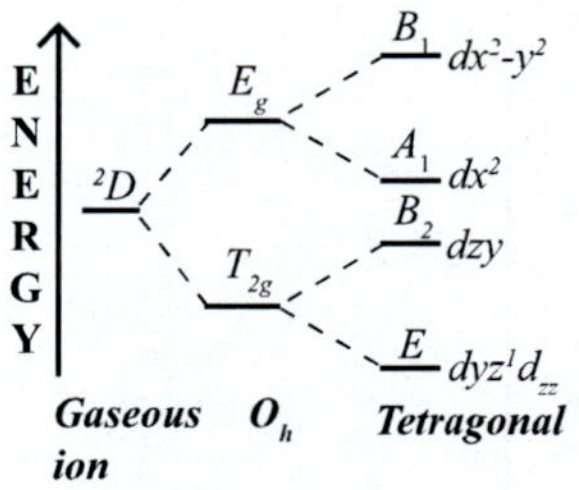

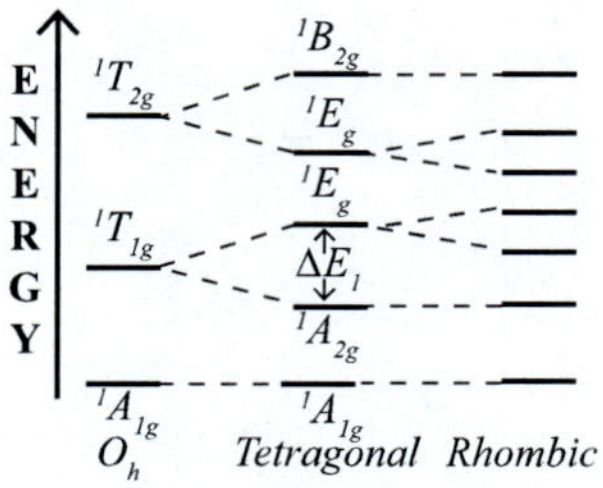

Figure 3.3 reports as example the electronic spectra of three Co(III) complexes with different symmetries (Fig. 3.3) (a) octahedral tris ethylenediamine (b) cis-difluoro bis ethylenediamine (c) trans-difluoro bis ethylenediamine. On passing from (a) to (b) and (c), the two bands gradually broaden, and at the end split in (c), the distortion from the octahedral symmetry is highest. More in general looking at the change of energy in the absorption spectra, as a function of the percent of distortion, Ni(II) complexes have the behavior in Scheme 3.9.

In general, large variations both in the number and in the energy of the states are observed by lowering the symmetry. Thus, it is not so easy to assign the symmetry and the electronic configuration in the presence of large distortion from the cubic symmetry. Instead it is sometimes simple to distinguish the tetrahedral from the octahedral symmetry of a given ion (Co(II) in Fig. 3.4), based on the intensity of transitions. These are much more intense in tetrahedral symmetry, due to the mixing between d and p orbitals allowed in T_d symmetry. By this mixing, the d-d transition is no more pure and no more forbidden.

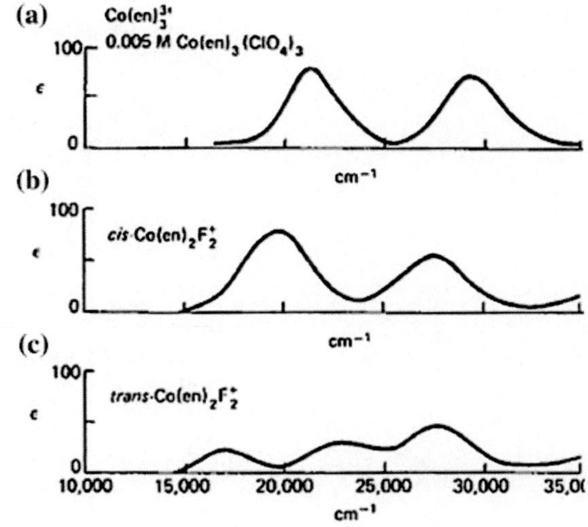

Fig. 3.3 Electronic spectra of Co(III) complexes ([1], p. 389)

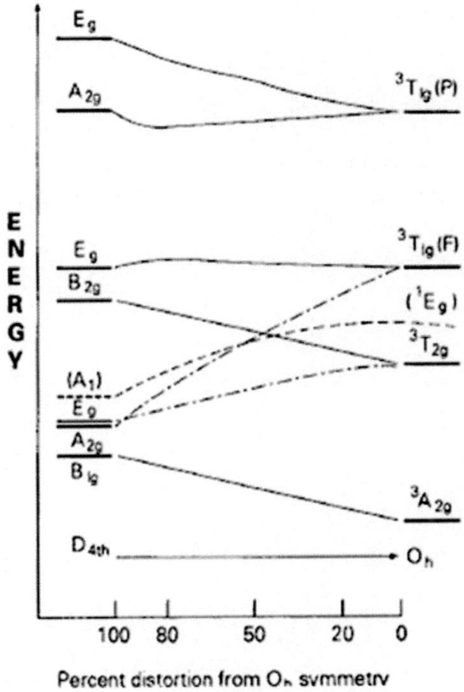

Scheme 3.9 Electronic spectra of Ni(II) complexes, as a function of the distortion from O_h

Fig. 3.4 Electronic spectra of octahedral and tetrahedral Co (III) complexes ([1], p. 392)

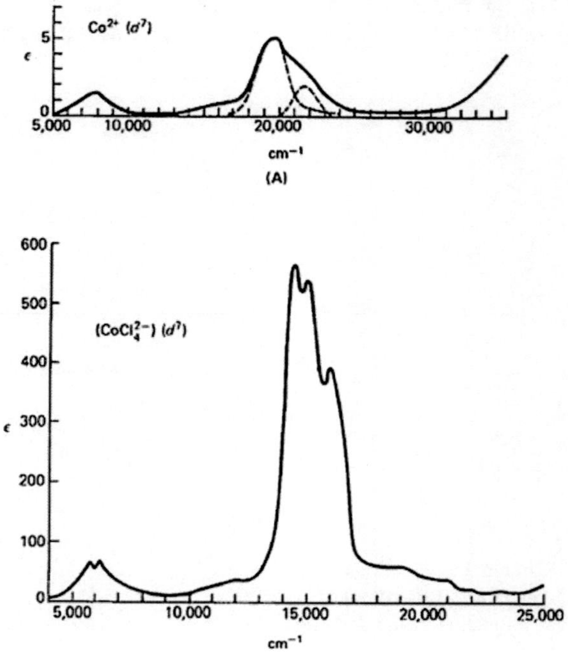

In conclusion, the group theory allows the prevision of the type of the electronic states due to the crystal field perturbation and of the expected transitions. By this approach, both the coordination geometry and the electronic configuration of the ion can be hypothesized.

Chapter 4
Magnetism

The magnetic properties play a fundamental role in the understanding of the electronic structure of ions and molecules. In fact starting from the energy of interaction between the electrons and the magnetic field, it becomes possible to locate the unpaired electrons of the ground state and to hypothesize the energy trend of the remaining electronic states. The magnetic data are obtained from measurements of magnetic susceptibility and from the electron spin resonance spectra.

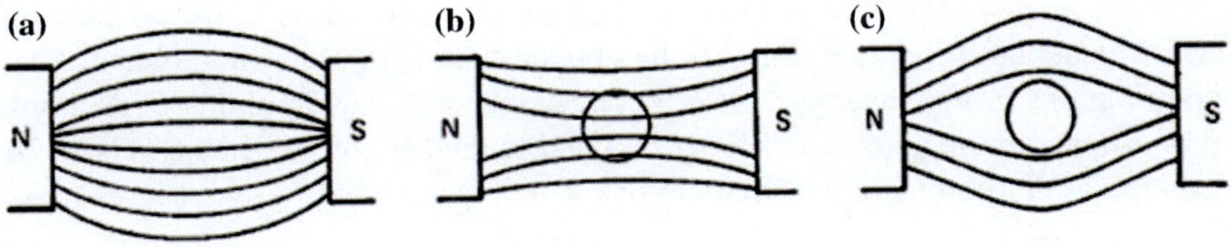

Different flux lines in vacuo (a), in the presence of paramagnetic material (b), in the presence of diamagnetic material (c)

In order to quantify this effect, it is convenient to define a quantity called magnetic induction expressed as vector $\vec{B}$

$$\vec{B} = \vec{H}_0 + 4\pi\vec{M}$$

$\vec{H}_0$ is the external field and $\vec{M}$ the intensity of magnetization per unit volume. Dividing by $\vec{H}_0$

$$\frac{B}{H_0} = 1 + 4\pi\frac{M}{H_0} = 1 + 4\pi\chi_r$$

F. Morazzoni, *Symmetry in Inorganic and Coordination Compounds*, Lecture Notes in Chemistry 106, https://doi.org/10.1007/978-3-030-62996-0_4

Table 4.1 Different types of magnetic behavior

Type	Sign	Magnitude	Field dependence of χ	Origin
Diamagnetism	–	10^{-6} emu units	Independent	Field induced, paired electron circulations
Paramagnetism	+	$0–10^{-4}$ emu units	Independent	Angular momentum of the electron
Ferromagnetism	+	$10^{-4}–10^{-2}$ emu units	Dependent	Spin alignment from dipole–dipole interaction of moments on adjacent atoms, ↑↓
Antiferromagnetism	+	$0–10^{-4}$ emu units	Dependent	Spin pairing, ↑↓, from dipole–dipole interactions

χ_v is the magnetic susceptibility for unit of volume, dimensionless

$\chi_v/d = \chi_g$ gram susceptibility (cm^3/gr) $\chi_g \cdot$ MW (cm^3/mole) is the molar susceptibility Table 4.1 and Scheme 4.1.

For our purposes, it is mainly important the origin of the magnetic behavior, as it is this aspect that can be relatable to the electronic properties. The variation of the magnetic properties with the temperature is also indicative of the different behavior. Consider that the diamagnetic behavior is independent of the temperature.

Diamagnetism arises from field-induced electron circulation of paired electrons, which generates a field opposed to the applied field. Thus, all the molecules have contributions from diamagnetic effects. The diamagnetic susceptibility of an atom is proportional to the number of electrons n and to the sum of the square value of the average orbital radius of the ith electron

$$\chi_A = -\frac{Ne^2}{6mc^2}\sum_{l}^{n}\overline{r_i^2} = -2.83 \times 10^{10}\sum_{i}^{n}\overline{r_i^2}$$

Scheme 4.1 Variation of the magnetic properties with temperature T_N, where T_N is the Néel temperature, and T_C is the Curie temperature

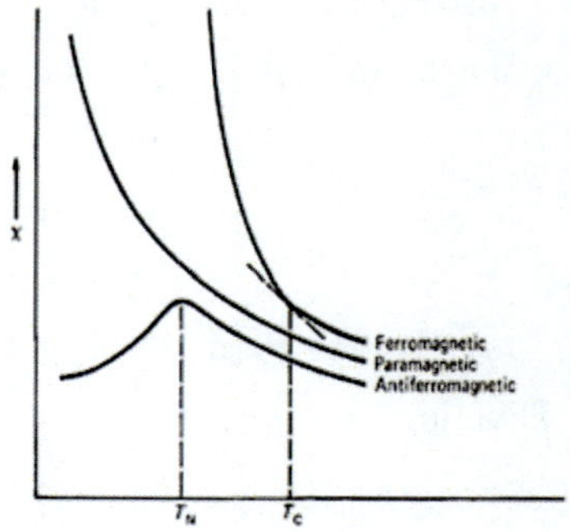

The molar diamagnetic susceptibility can be obtained by the sum of the diamagnetic contributions of all the atoms χ_A and of the functional groups χ_B, as indicated in Table 4.2. When a magnetic measurement is performed, the magnetic susceptibility is the difference between the measured value and the value of the diamagnetic susceptibility.

$$\chi_{PARA} = \chi_{MEAS} - \chi_{DIA}$$

An example, the pyridine molecule

C_5H_5N

SUM OF CONTRIBUTIONS TO χ ($\times$ 10^{-6} CM^3 MOLE^{-1})

$$\begin{aligned} 5 \times \mathrm{C\,(ring)} &= -31.2 \\ 5 \times \mathrm{H} &= -14.6 \\ 1 \times \mathrm{N\,(ring)} &= -4.6 \\ \chi = \sum_i \chi_{A_i} + \sum_j \chi_{B_j} &= -50.4 \times 10^{-6}\,\mathrm{cm}^3\,\mathrm{mole}^{-1} \end{aligned}$$

The functional groups are accounted for by using the ring values for carbon and nitrogen, so $\Sigma\chi_r$ equals zero.

The calculation of the diamagnetic contribution allows, after the experimental measure of the total magnetic contribution, the determination of the paramagnetic contribution. The goal is at this point to relate the value of the paramagnetic contribution to the electronic properties.

4.1 Interaction Between Electrons and Magnetic Field

The effect of the magnetic field on the electronic spin is better explainable in terms of the quantum mechanical approach than by the magnetic field classical interaction. It is well known that each system having discrete energy levels may be associated to the (4.1)

$$\Lambda_i \psi_i = \lambda_i \psi_i \tag{4.1}$$

where Λ is the operator describing the potential energy of the system to which we associate the energy states i, eigenvalues, and the wave functions i, eigenfunctions. The Λ determination is a delicate point, to be done by attempts.

In this case, we will proceed through the following considerations:

(a) By immersing a magnetic dipole μ in a magnetic field H (Fig. 4.5), the energy of the system is given by the expression (4.2)

Table 4.2 Diamagnetic contributions

Atoms, χ_A				Bonds, χ_B	
Atom	$\chi_A(\times 10^{-6}cm^3\ mole^{-1})$	Atom	$\chi_A(\times 10^{-6}cm^3\ mole^{-1})$	Bond	$\chi_B(\times 10^{-6}cm^3\ mole^{-1})$
H	−2.93	F	−63	C=C	+5.5
C	−6.00	CI	−20.1	C≡C	+0.8
C (aromatic)	−6.24	Br	−30.6	C=N	+8.2
N	−5.57	1	−44.6	C≡N	+0.8
N (aromatic)	−4.61	Mg^{2+}	−5	N=N	+1.8
N (monamide)	−1.54	Zn^{2+}	−15	N=O	+1.7
N (diamide, imide)	−2.11	Pb^{2+}	−32.0	C=O	+6.3
O	−4.61	Ca^{2+}	−10.4		
O_2 (carbosylate)	−7.95	Fe^{2+}	−12.8		
S	−15.0	Cu^{2+}	−12.8		
P	−26.3	Co^{2+}	−12.8		
		Ni^{2+}	−12.8		

$$W = -\vec{\mu} \cdot \vec{H} = -\vec{\mu} \cdot H \cos(\vec{\mu} H) \quad (4.2)$$

(b) Both the orbital and the spin angular momentum undergo quantization (Figs. 4.1 and 4.2), as well as their components in a given space direction, being their values

$$L(L+1)^{1/2} h/2\pi \quad S(S+1)^{1/2} h/2\pi$$

M_L 2L+1 values from –L to +L
M_S 2S+1 values from –S to +S

(c) The moment of the magnetic dipole is proportional both to the orbital and the spin angular momentum.

The property associated to the rotation of the electron negative charge, and the related relations are

$$\mu_z = \gamma M_L h/2\pi$$
$$\mu_z = \gamma M_s h/2\pi \quad \textit{and if } \gamma = -ge/2mc \quad \beta = eh/4\pi mc \rightarrow$$
$$\mu_z = -g\beta M_S$$

From the a and c considerations, the energy of one electron with $M_S = \pm \frac{1}{2}$ is

$$W = g\beta H M_S = \pm 1/2 g\beta H \quad \Delta = g\beta H$$

Based on points a–c, the spin properties of the system under investigation determine its behavior; thus, the Hamiltonian equation for a spin only case results in

$$S_z\psi_i = M_s\psi_i \quad S_z\psi_i(M_s \pm 1/2) = \pm 1/2 M_s\psi_i(M_s \pm 1/2)$$

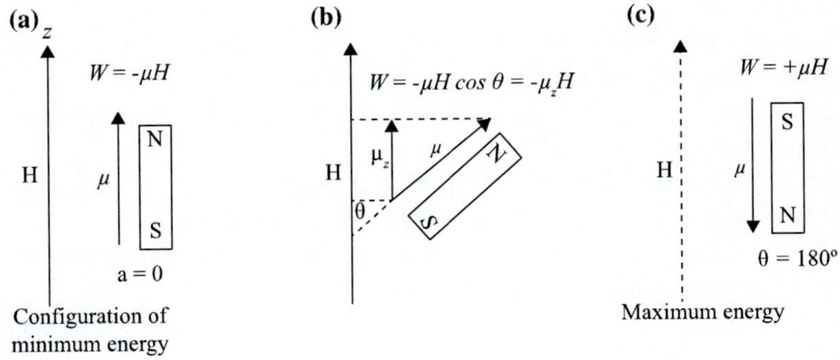

Fig. 4.1 Energy interaction of one electron in a magnetic field ([2], pp. 8, 11, 13)

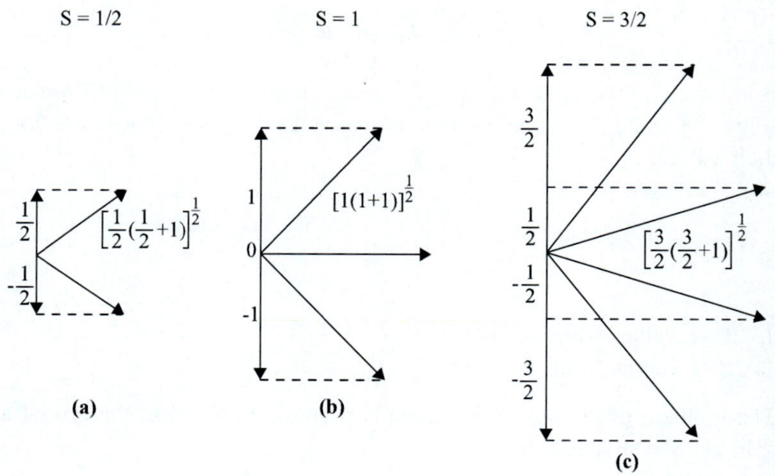

Fig. 4.2 Quantum mechanical effect on the spin angular momentum ([2], pp. 8, 11, 13)

As S_z is commutative with the Hamiltonian operator H, $H\psi_i(M_s \pm 1/2) = W_{\pm 1/2}\psi_i(M_s \pm 1/2)$

By associating to the magnetic moment μ_z, the corresponding operator

$$\mu_z = -g\beta M_s$$
$$\mu_z = -g\beta S_z$$
$$\textit{and to the energy} \quad W = -\mu_z H$$
$$\textit{The corresponding Hamiltonian is} \quad H = g\beta H S_z$$
$$W_{1/2} = 1/2 g\beta H$$
$$W_{-1/2} = -1/2 g\beta H$$
$$\Delta W = g\beta H$$

See Fig. 4.3.

The energy transition is ΔW *and is tuned by* H, provided $\Delta M_s = \pm 1$ as selection rule g is a value peculiar of the electron behavior.

See in Figs. 4.1, 4.2, and 4.3 the comparison between quantum mechanical and classical approaches:

Energy of a classical magnetic dipole in a magnetic field as a function of the angle Θ between the magnetic field and the axis of the dipole.

Allowed values of the total spin angular $[S(S+1)]^{1/2}$ moment and of the component M_s (in units h/2π) in a fixed direction.

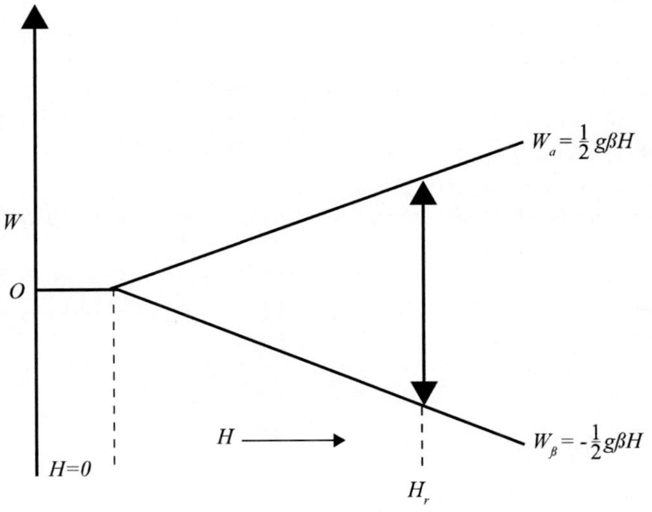

Fig. 4.3 Energy changes induced by the magnetic field on one electron ([2], pp. 8, 11, 13)

4.2 Magnetic Susceptibility Expression for S = 1/2

Considering that for a given electron n $\mu_n = -g\beta\, M_S$ and $W_n = g\,\beta\, H\, M_S = \pm\, ½\, g\, \beta$ H and that the energy difference between the spin states is lower than kT, both the states of S = 1/2 are populated at room temperature. The probability P_n for populated states with energy E_n is given by

$$P_n = \frac{N_n}{N} = \frac{\exp\left(\frac{-E_n}{kT}\right)}{\sum\limits_n \exp\left(\frac{-E_n}{kT}\right)}$$

N_n refers to the population of the state n, while N to the total population of all the existing states. The population-weighted sum of magnetic moments over the individual states gives the macroscopic magnetic moment M. For a mole of material

$$M = \underline{N} \sum_{m_s} \mu_n P_n$$

Where N is the number of Avogadro. Substituting P_n

$$M = \frac{\underline{N} \sum\limits_{m_s=-1/2}^{+1/2} \mu_n \exp\left(\frac{-E_n}{kT}\right)}{\sum\limits_{m_s=-1/2}^{+1/2} \exp\left(\frac{-E_n}{kT}\right)}$$

Substituting μ_n ed E_n

$$M = \frac{Ng\beta}{2}\left[\frac{\exp\left(\frac{g\beta H}{2kT}\right) - \exp\left(\frac{-g\beta H}{2kT}\right)}{\exp\left(\frac{g\beta H}{2kT}\right) + \exp\left(\frac{-g\beta H}{2kT}\right)}\right]$$

in general given the short distance between the magnetic levels $g\beta H/KT \ll 1$, thus

$$\exp\left(\frac{\pm g\beta H}{2kT}\right) \approx \left(1 \pm \frac{g\beta H}{2kT}\right)$$

$$M = \frac{Ng^2\beta^2 H}{4kT}$$

since $\chi_{para} = \vec{M}\vec{H}$

$$\chi = \frac{Ng^2\beta^2}{4kT} \quad \chi = \frac{Ng^2\beta^2}{3kT}S(S+1)$$

This expression is the **Curie law**. This expression gives the paramagnetic susceptibility for spin only systems

If we define a new scalar quantity

$$\mu_{\text{eff}} = \left(\frac{3k}{N\beta^2}\right)^{1/2}(\chi T)^{1/2} = 2.828(\chi T)^{1/2}(\text{BM})$$

$$\mu_{\text{eff}}(\text{ spin - only }) = g[S(S+1)]^{1/2}(\text{BM})$$

Some examples are here reported

Number of unpaired electrons	S	μ_{eff}(spin - only)(BM)
1	1/2	1.73
2	1	2.83
3	3/2	3.87
4	2	4.90
5	5/2	5.92
6	3	6.93
7	7/2	7.94

Thus, the μ_{eff} value allows to detect the number of unpaired electrons in spin only systems. If the systems include magnetic contribution different from that of the spin one, the μ_{eff} values are different. In this case, the system is associated to the Hamiltonian

$$\hat{H} = \lambda\hat{L}\cdot\hat{S} + \beta(\hat{L} + g_e\hat{S})\cdot H$$

$$E_{\mathrm{n}} = \underset{\lambda \mathrm{L}\cdot S}{E_{\mathrm{n}}^{(0)}} + \underset{\text{first-order Zeeman (diagonal terms)}}{HE_{\mathrm{n}}^{(1)}} + \underset{\text{sccond-order Zeeman (off-diagonal terms)}}{H^2E_{\mathrm{n}}^{(2)}}$$

By replacing in the Curie expression exp(–E_n/kT) with

$$\exp\left(\frac{-E_{\mathrm{n}}^{(0)} - HE_{\mathrm{n}}^{(1)} - H^2E_{\mathrm{n}}^{(2)} + \cdots}{kT}\right) \cong \left(1 - \frac{HE_{\mathrm{n}}^{(1)}}{kT}\right)\exp\left(\frac{-E_{\mathrm{n}}^{(0)}}{kT}\right)$$

$$\mu_{\mathrm{n}} = \frac{-\partial E_{\mathrm{n}}}{\partial H} = -E_{\mathrm{n}}^{(1)} - 2HE_{\mathrm{n}}^{(2)}$$

$$M = \underline{N}\frac{\sum_n \left(-E_{\mathrm{n}}^{(1)} - 2HE_{\mathrm{n}}^{(2)}\right)\left(1 - \frac{HE_{\mathrm{n}}^{(1)}}{kT}\right)\exp\left(\frac{-E_{\mathrm{n}}^{(0)}}{kT}\right)}{\sum_n \exp\left(\frac{-E_{\mathrm{n}}^{(0)}}{kT}\right)\left(1 - \frac{HE_{\mathrm{n}}^{(1)}}{kT}\right)}$$

Being M = 0 for H = 0

$$-\sum_n E_{\mathrm{n}}^{(1)}\exp\left(\frac{-E_{\mathrm{n}}^{(0)}}{kT}\right) = 0$$

Neglecting the terms higher than $E_{\mathrm{n}}^{(2)}$ and $E_{\mathrm{n}}^{(2)} \times E_{\mathrm{n}}^{(1)}$

$$\chi = \underline{N}\frac{\sum_n \left[\frac{\left(E_{\mathrm{n}}^{(1)}\right)^2}{kT} - 2E_{\mathrm{n}}^{(2)}\right]\exp\left(\frac{-E_{\mathrm{n}}^{(0)}}{kT}\right)}{\sum_n \exp\left(\frac{-E_{\mathrm{n}}^{(0)}}{kT}\right)}$$

This is the *Van Vleck equation*, where the magnetic susceptibility contains different contributions to the interaction with the magnetic field, that field independent and that first order dependent. The difference of μ_{eff} with respect to the spin only value is due to the contribution of the orbital magnetic moment. It is mainly this difference to contribute the assignment of the electronic state. If an electron can occupy degenerate orbitals that permit circulation of the electron about an axis, an orbital angular momentum can result. Table 4.3 reports the cases where the angular momentum is active or quenched.

The behavior depends on the electronic configuration, on the crystal field symmetry, and thus, the quenching or not gives an indication of the coordination geometry and of the electronic configuration.

Let us consider the d^3 electronic configuration: In octahedral field symmetry, the three electrons lie in t_{2g} orbitals, with parallel spins, and no orbital circulation is

Table 4.3 Relations between magnetism and crystal field properties for different d^n

		Stereochemistry					
		Octahedral			Tetrahedral		
No. of d-electrons	Free ion Ground term	$t_{2g}^{n}e_{g}^{m}$ ground configuration	Ligand field ground term	Quenching of orbital contribution	Ground configuration	Ligand field ground term	Quenching of orbital contribution
1	^{2}D	$t_{2}^{1}g$	$^{2}T_{2g}$	No	e^{1}	^{2}E	Yes
2	^{3}F	$t_{2}^{2}g$	$^{3}T_{1g}$	No	e^{1}	$^{3}A_{2}$	Yes
	^{4}F	$t_{2}^{3}g$	$^{4}A_{2g}$	Yes	$e^{2}t_{2}^{1}$	$^{4}T_{1}$	No
4	^{5}D	$t_{2}^{3}g^{e}{}_{g}^{1}$	$^{5}E_{g}$	Yes	$e^{2}t_{2}^{2}$	$^{5}T_{2}$	No
		$t_{2}^{4}g$	$^{3}T_{1g}$	No	–	–	–
5	^{6}S	$t_{2}^{3}g_{g}^{2}$	$^{6}A_{1g}$	Yes	$e^{2}t_{2}^{3}$	$^{6}A_{1}$	Yes
		$t_{2}^{5}g$	$^{2}T_{2g}$	No	–	–	–
6	^{5}D	$t_{2}^{4}g_{g}^{2}$	$^{5}T_{2g}$	No	$e^{3}t_{2}^{3}$	^{5}E	Yes
		$t_{2}^{6}g$	$^{1}A_{1g}$	Yes	–	–	–
7	^{4}F	$t_{2}^{5}g_{g}^{2}$	$^{4}T_{1g}$	No	$e^{4}t_{2}^{3}$	–	Yes
		$t_{2}^{6}g^{-1}$	$^{2}E_{g}$	Yes	–	–	–
8	^{3}F	$t_{2}^{6}g_{g}^{2}$	$^{3}A_{2}g$	Yes	$e^{4}t_{2}^{4}$	$^{3}T_{1}$	No
9	^{2}D	$t_{2}^{6}e_{g}^{3}$	$^{2}E_{g}$	Yes	$e^{4}t_{2}^{5}$	$^{2}T_{2}$	No

allowed. Thus, the quenching of the orbital momentum is fully justified. Instead in tetrahedral symmetry field, there is one electron in t_{2g} orbitals, and the orbital circulation is active.

4.3 Electron Spin Resonance

The resonance condition (energy of the magnetic transition) for one electron is $W = g\,\beta\,H\,M_S = \pm\,½\,g\,\beta\,H\,\Delta = g\,\beta\,H$ resonance condition $g = h\nu/\beta H_r$ where ν is the irradiation frequency and H_r is the magnetic field where the absorption happens (Fig. 4.4).

The experimental choice is to fix the frequency ν and to measure the field H_r. The g value depends on the electronic surrounding of the unpaired electron, and thus, it is typical of a given paramagnetic center. In the case of a free electron, g = 2.0023. Values are different due to spin–orbit coupling interaction and can induce anisotropic behavior (g is represented by a tensor) in the form of cubic, axial, or rhombic symmetry. In order to carefully measure the g value, it is mandatory to precisely measure

$$\nu\, and\, H_r$$

In general, the measure is obtained by using a precisely measured ν and calibrating the magnetic field by a standard sample with known g value

$$g_s = h\nu/\beta H_s \quad g_x = h\nu/\beta H_x \quad g_s/g_x = H_x/H_s$$

A used standard molecule is diphenyl pycrilhydrazil $(DPPH)g = 2.0037$.

Block diagram of a typical X-band ESR spectrometer employing 100 gHz phases-sensitive detection is as it follows (Fig. 4.5). The shape of the spectrum corresponds to Gaussian or Lorenzian lines (Fig. 4.6) and is taken mostly as a derivative curve, to identify the overlapping absorptions. The most common used frequencies of the microwave source are 9.417 GHz (X band) 35 GHz (Q band)

Figure 4.7 shows the real spectrometer.

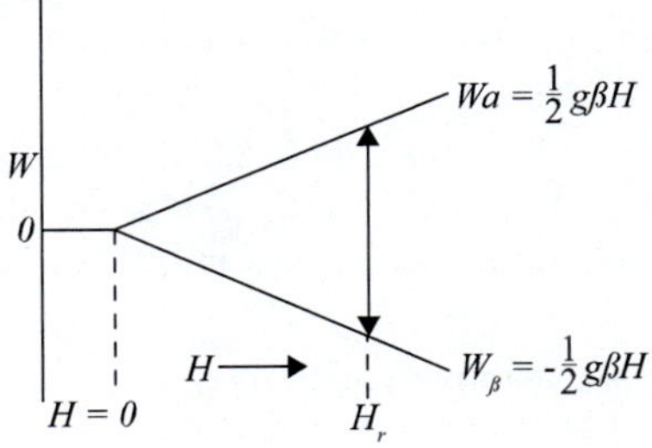

Fig. 4.4 Energy changes induced by the magnetic field on one electron ([2], p. 24)

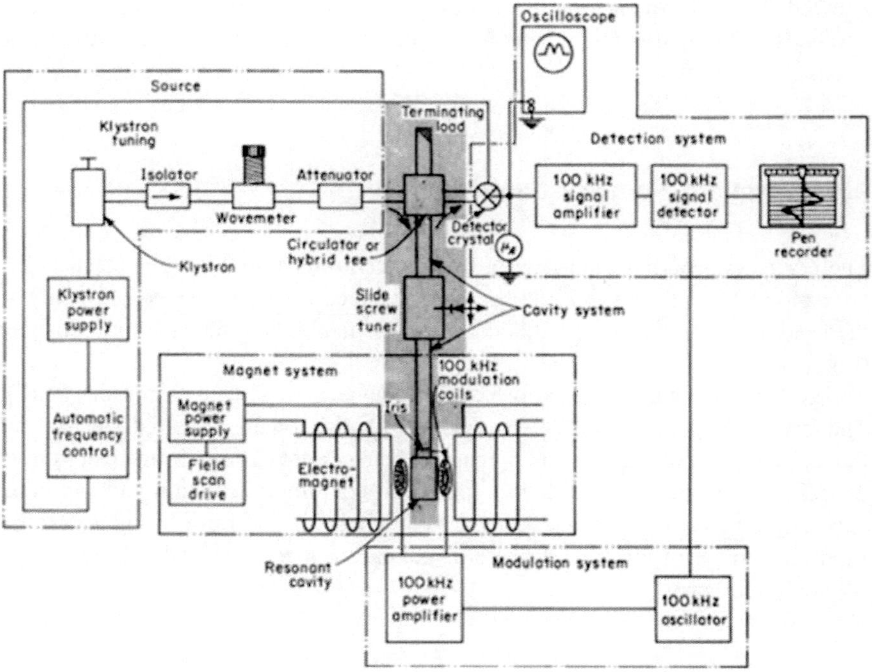

Fig. 4.5 Scheme of the ESR apparatus ([2], p. 34)

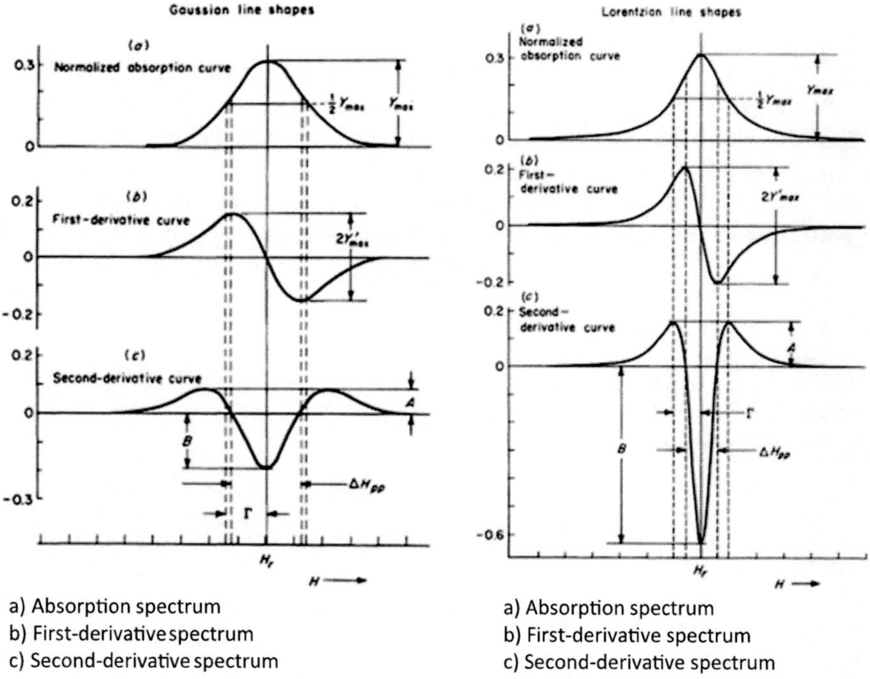

Fig. 4.6 Spectra general shape ([2], pp. 157 e 158)

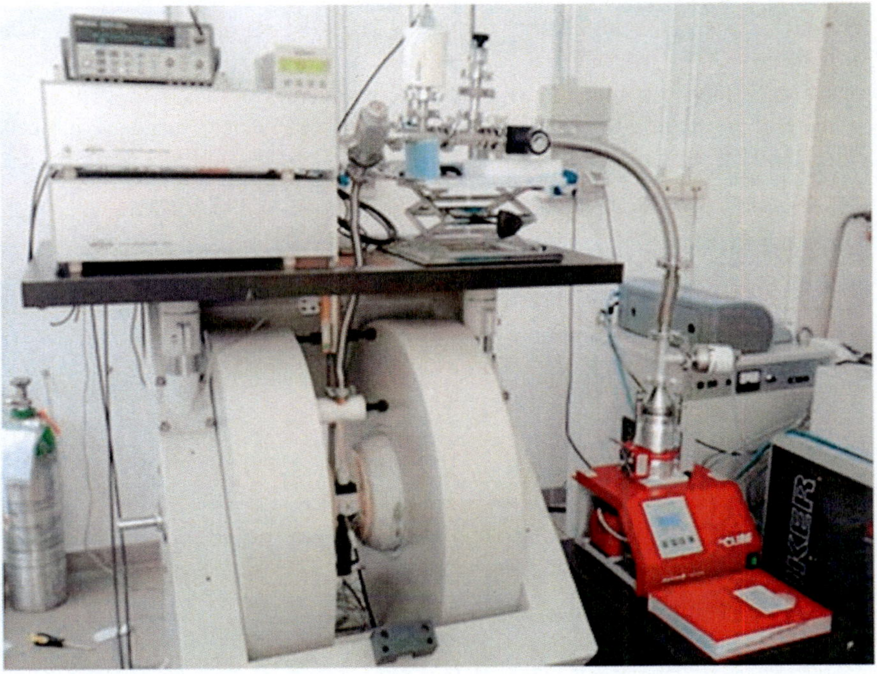

Fig. 4.7 Electron spin resonance spectrometer

4.3.1 Hyperfine Nuclear Interaction

The electron magnetic moment and the nuclear magnetic moment interact, modifying the microwave absorption. The number of single resonances increases, although they are centered on a unique resonance field H_r which corresponds to the absorption field in the absence of electron–nucleus interaction.

$$H_{eff} = H + H_{local}$$

H_{local} is due to nuclear interaction. From the quantum mechanical approach for the spin nuclear momentum I, 2I + 1 components are expected. In the case of hydrogen atom I = ½ M_I = ± ½, the spectrum shows two resonances.

$$H_{eff} = H \pm a/2 = H \pm aM_i \, (where\, a\, is\, a\, measure\, of\, the\, interaction)$$

The energy of hyperfine interaction in the isotropic form was investigated by Fermi and is defined with the following equation.

$$W_{iso} = 8\pi/3(\psi_0)^2 \mu_{ez}\mu_{nz}$$

where ψ_0 is the wave function evaluated on the nucleus.

μ_{ez} and μ_{nz} are the components of electronic and nuclear magnetic moments along the H direction.

4.3.2 *Hyperfine Interaction by Quantum Mechanical Approach*

By using quantum mechanical operators:

$$M_{ez} = -g\beta S_z$$
$$\mu_{nz} = -g_n\beta_n I_z$$
$$H = 8\pi/3 g\beta g_n\beta_n(\psi_0)^2 S_z I_z = hA_0 S_z I_z$$

where A_0 is the isotropic hyperfine coupling constant (hertz). In the case of the hydrogen atom, the complete Hamiltonian operator is

$$H = g\beta HS_z + hA_0 S_z I_z$$

By associating the eigenvalues of S_z and of I_z to the four spin and nuclear states

$$M_S = 1/2 \equiv \alpha_e \, M_S = -1/2 \equiv \beta_e$$
$$M_I = 1/2 \equiv \alpha_N \, M_I = -1/2\beta_N$$

four different states are generated $(\alpha_e\alpha_N)(\beta_e\alpha_N)(\alpha_e\beta_N)(\beta_e\beta_N)$

Operating by the H operator on each one of the states

$$W_{\alpha e\,\alpha N} = <\alpha_e\alpha_N|g\beta HS_z + hA_0 S_z I_z|_e\alpha_N> \; = 1/2g\beta H + 1/4\,hA_0$$

$$W_{\alpha e\beta N} = 1/2g\beta H - 1/4hA_0$$
$$W_{\beta e\beta N} = -1/2g\beta H + 1/4hA_0$$
$$W_{\beta e\alpha N} = -1/2g\beta H - 1/4hA_0$$

The microwave-induced energy transition undergoes the following selection rules

$$\Delta M_s \pm 1\Delta M_I = 0$$

The energy transition is consequently

Scheme 4.2 Effect of the hyperfine interaction for hydrogen atom

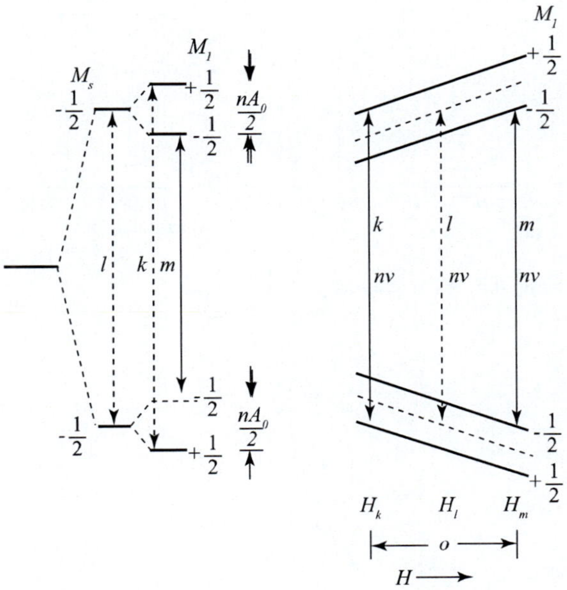

$$H_k = h\nu_0/g\beta - a/2 \quad M_I = 1/2$$
$$H_m = h\nu_0/g\beta + a/2 \quad M_I = -1/2$$

where $a = hA_0/g\beta$ (Scheme 4.2)

4.3.3 The Electronic Interaction with the Magnetic Field in Oriented Systems

In the most general case, the interaction energy of an electron with the magnetic field is anisotropic (Figs. 4.8 and 4.9), provided the matter physical status does not induce averaging of the magnetic interactions. g is represented by a tensor, and the Hamiltonian operator is

$$H = \beta SgH \quad S \; and \; H \; are \; both \; vectors$$
$$H = \beta\left(S_x g_{xx} H_x + S_y g_{yy} H_y + S_z g_{zz} H_z\right)$$

4.3.4 The Electronic Interaction with the Magnetic Field in Oriented Systems

g_{xx}, g_{yy}, and g_{zz} are the g components along three directions, called principal magnetic axes, observed when the magnetic field is oriented along one of these directions.

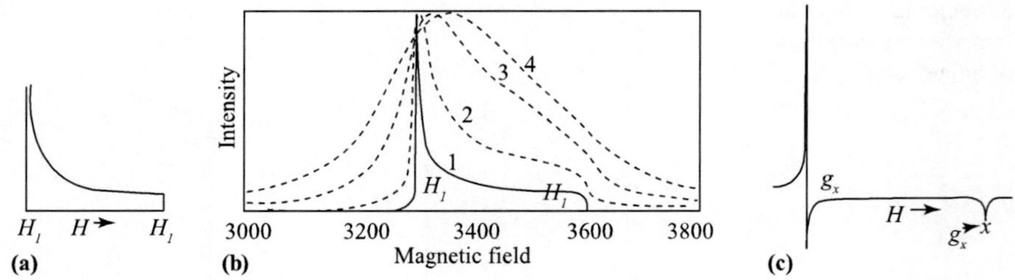

Fig. 4.8 Axial symmetry

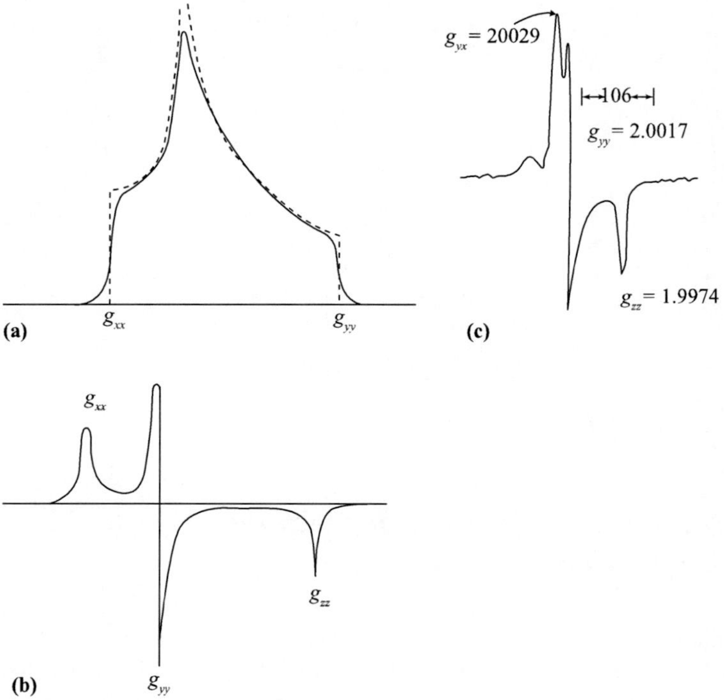

Fig. 4.9 Resonance spectra for rhombic symmetry ([2], pp. 157 e 158)

For different orientations, the g value follows the expression:

$$g_{\text{eff}}^2 = g_{\text{xx}} cos^2 \vartheta_{\text{HX}} + g_{\text{yy}} cos^2 \vartheta_{\text{HY}} + g_{\text{zz}} cos^2 \vartheta_{\text{HZ}}$$

ϑ is the angle between the magnetic field component and the g tensor component.

In the following $\cos^2\vartheta_{HZ} = l_z$

$$g^2_{\text{eff}} = [l_x l_y l_z] \begin{matrix} g^2_{xx} & 0 & 0 \\ 0 & g^2_{yy} & 0 \\ 0 & 0 & g^2_{yy} \end{matrix} \begin{matrix} [l_x \\ l_y \\ l_z] \end{matrix}$$

The g_{eff}^2 matrix is diagonal only when the H directions are coincident with the magnetic ones (principal directions).

In general, the principal magnetic directions are unknown, and g is measured with respect to an arbitrary system, that is

$$g^2_{\text{eff}} = [l_x l_y l_z] \begin{matrix} g^2_{xx} & g^2_{xy} & g^2_{xz} \\ g^2_{yx} & g^2_{yy} & g^2_{yz} \\ g^2_{zx} & g^2_{zy} & g^2_{zz} \end{matrix} \begin{matrix} [l_x \\ l_y \\ l_z] \end{matrix}$$

The matrix is then diagonalized to obtain g principal values and their orientation with respect to the magnetic field.

4.3.5 The Origin of the g Anisotropic Behavior

The observation that the **g** parameter has anisotropic behavior is not simply relatable to the symmetry of the field where the electron is located and interacts with the external magnetic field. It is more suitably explained as the interaction between the external magnetic field and both the spin and the orbital electron magnetic moments. The total interaction is described by the Zeeman operator:

$$H = \beta HL + g_e \beta HS$$

If the spin and the orbital angular moments interact between them, the spin–orbit coupling operator ζ L S perturbs the spin eigenfunctions

$$\psi_0|\alpha/\beta > \textit{ to new states}|$$
$$\pm > \; = |\psi_0\alpha/\beta > \; - \Sigma_n <n|\zeta LS|\psi_0\alpha/\beta > \; \cdot (1/E_n - E_0)|n >$$

Let us consider a new operator $\underline{S}$ which operates as it follows:

$$\underline{S}_z|\pm > \; = \pm 1/2|\pm > \quad \underline{S}_x|+ > \; = 1/2|- > \quad <\underline{S}_y|+ > \; = 1/2i|- >$$

where $\underline{S}$ works on the states |±> as S on the states |α/β>

Then, if the magnetic field is along the z-axis, the Hamiltonian operator is

$H = \beta H\left(g_{xz}\underline{S}_x + g_{yz}\underline{S}_y + g_{zz}\underline{S}_z\right)$, and it operates on the $|\pm>$ eigenfunctions as it follows

$$\begin{matrix} <+|1/2\beta H g_{zz} & 1/2\beta H\left(g_{xz} - ig_{yz}\right) \\ <-|1/2\beta H\left(g_{xz} + ig_{yz}\right) & -1/2\beta H g_{zz} \\ <+| & <-| \end{matrix} \qquad (a)$$

The true Zeeman Hamiltonian for H along z is $H = \beta H(L_z + g_e S_z)$, and it is associated to the energy matrix

$$\begin{matrix} \beta H < +|L_z + g_e S_z| + > & \beta H < +|L_z + g_e S_z| - > \\ \beta H < -|L_z + g_e S_z| + > & \beta H < -|L_z + g_e S_z| - > \end{matrix} \qquad (b)$$

Considering the analogy between matrices *(a)* and *(b)*, the g component expression can be as it follows

$$g_{zz} = 2 < +|L_z + g_e S_z| + > = g_e - 2\zeta\Sigma_n < \psi_0|L_z|\psi_n > < \psi_n|L_z|$$
$$\psi_0 > \cdot 1/(E_n - E_0)$$
$$g_{xz} + ig_{vz} = 2 < -|L_z + g_e S_z| + > = g_e - 2\zeta\Sigma_n < \psi_0|L_z|\psi_n > < \psi_n|L_x|$$
$$\psi_0 > \cdot 1/(E_n - E_0)$$

The g values contain the free electron value, g_e, 2.0023, and a perturbation term which is a function of the spin–orbit coupling and of the energy difference between the ground electronic state and the excited ones interacting by spin–orbit coupling.

4.3.6 g Expressions for a d^1 Electronic Configuration in Tetragonal Symmetry Field

$$d_{(x2-y2)} < d_{xz,yz} < d_{z^2}$$

If the unpaired electron lies in the d $x^2 - y^2$ orbital,

$$g_{zz} = 2 - 8\zeta/\Delta \quad \Delta = E\left(d_{x2-y2}\right) - E\left(d_{xy}\right)$$
$$g_{xx,yy} = 2 - 2\zeta/\delta \quad \delta = E\left(d_{xz,yz}\right) - E\left(d_{x2-y2}\right)$$

If the unpaired electron lies in the $d(z^2)$ energy of orbital,

$$g_{zz} = 2$$
$$g_{xx,yy} = 2 - 6\zeta/E\left(d_{xz,yz}\right) - E(d_{z^2})$$

If the unpaired electron lies in the (d_{xy}) orbital,

$$g_{zz} = 2 - 8\zeta/\Delta$$
$$g_{xx,yy} = 2 - 2\zeta/\delta$$

4.3.7 g Tensor Dependence on the Energy Level Trend of the Paramagnetic Centers

In the case of valence electrons

$$g_{\mathrm{ij}} = g_{\mathrm{e}} - 2\zeta \sum_{n=\psi_0} \frac{<\psi_0|L_{\mathrm{i}}|\psi_{\mathrm{n}}> <\psi_{\mathrm{n}}|L_{\mathrm{j}}|\psi_0>}{E_{\mathrm{n}}^0 - E_0^0}$$

In the case of gap electrons

$$g_{ij} = g_e - 2\zeta \sum_{n\neq\psi_0} \frac{<\psi_0|L_{\mathrm{i}}|\psi_{\mathrm{n}}> <\psi_{\mathrm{n}}|L_{\mathrm{j}}|\psi_0>}{E_{\mathrm{n}}^0 - E_0^0}$$

E_0 near C_B g $<g_e$ E_0 near V_B g$>$ g_e

Let us observe that the expression indicates the free electron g value, perturbed by the contribution of the angular momentum, weighted on the energy differences among the interacting states.

Fig. 4.10 shows some examples of ESR spectra in different physical states.

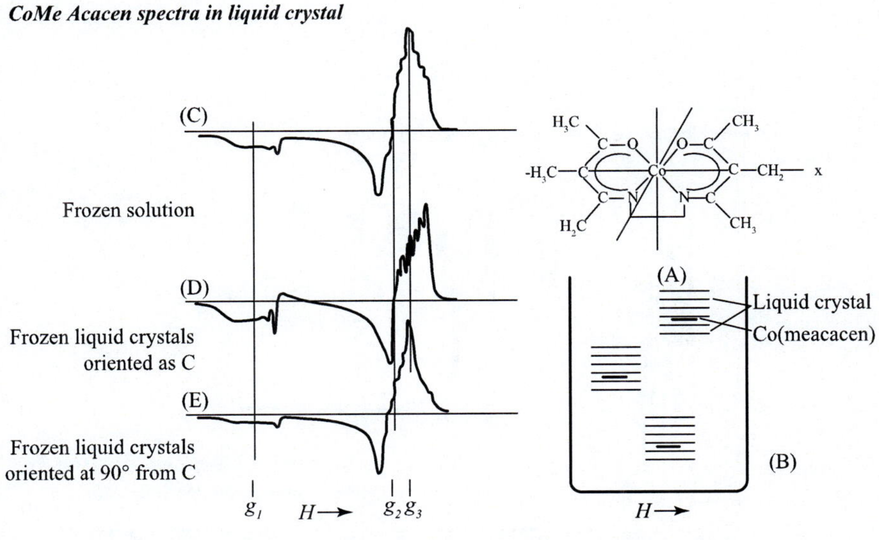

Fig. 4.10 Some examples of electron spin resonance spectra in different physical states

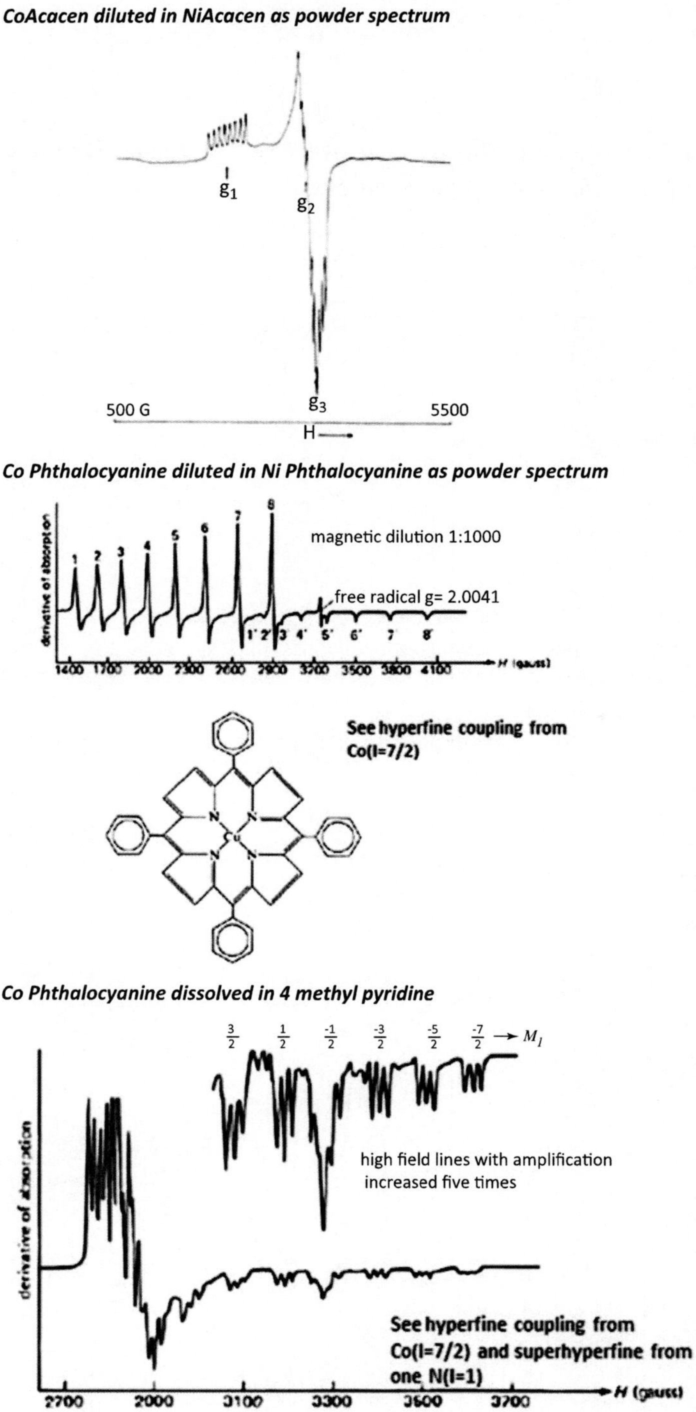

Fig. 4.10 (continued)

General remarks

The reported lectures outline the fundamental role of the group theory in simplifying the modelling of the chemical bond. It is showed the possibility of fast interpretation of the results of electronic spectroscopy and magnetism.

A simple quantum mechanical treatment is used, being the wave equations resolved by the aid of the symmetry operators.

For a chemist, it is a great value to simplify the quantum mechanical approach, as the molecules are complex systems whose electronic properties risk to be absolutely unknown.

For me, there was no better satisfaction as to look to a spectrum and in a short time being able to approach the molecular structure.

The applications which follow will demonstrate it.

The examples have been selected on the basis they apply the principles described before and clearly demonstrate that the results of the investigation help to understand the functionality of the compounds.

Chapter 5
The Symmetry Properties Describe the Electronic Structure of Inorganic and Coordination Compounds

The previous four chapters reported the fundamentals of the symmetry properties with the use of the group theory, aiming to describe the electronic structure of transition metal complexes and to make prevision of it. These principles are important not only as a base knowledge of the matter structure, but as a starting point to predict reactivity. Reactivity is in turn essential to explain the functionality of the chemical units.

The symmetry-based methods will be used in the following to show which way the case studies can be resolved. Symmetry, electronic structure, and magnetic properties will contribute together. The examples have been taken from published results.

The description will show the case, then it will report the discussion of the case, and finally it will give a functional chemical message.

5.1 The Case of $[Fe_3 Pt_3 (CO)_{15}]^{n-}$ n = 2, 1, and of $[Fe_3 Pt_3 (CO)_{15}]^{2\text{-}n}$

These organometallic compounds belong to the carbonyl cluster family and are constituted by Fe and Pt centers linked one to another and to terminal CO groups, as in Fig. 5.1 [3,4]. Several electronic and structural problems are connected to carbonyl clusters and more in general to metal clusters, beginning from the model of chemical bond between the metal centers and going to hypotheses of electronic properties connected with their reactivity. This should in principle allow to understand how the metal centers undergo interaction and increase the cluster nuclearity (number of interacting metal centers). The crystal and ligand field theories are not suitable to rationalize the bond interaction as any single metal center is not coordinated with only CO ligand molecules. Basing on the symmetry properties of the molecules in Fig. 5.1, belonging to D_{3h} group, a description can be

F. Morazzoni, *Symmetry in Inorganic and Coordination Compounds*, Lecture Notes in Chemistry 106, https://doi.org/10.1007/978-3-030-62996-0_5

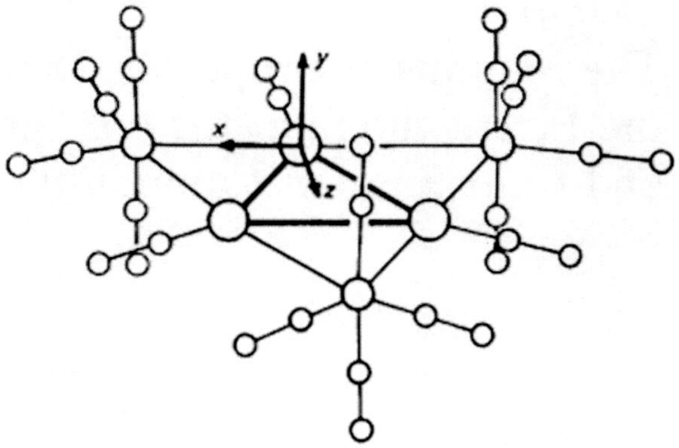

Fig. 5.1 Schematic representation of the structure of $[Fe_3\,Pt_3\,(CO)_{15}]^{n-}$ n = 1, 2 [3]

attempted. A local right-handed coordinate system for each platinum atom, with the z-axis directed toward the center of the Pt_3 triangle, the x-axis in the plane of Pt atoms, and the y-axis perpendicular to this plane, may be chosen. The nine valence platinum atomic orbitals, 6s $6p_x$ $6p_z$ of three Pt, are used for σ bond to three terminal carbonyls and three $Fe(CO)_4$ groups, and their bonding combinations with suitable ligand orbitals are filled with 18 of the 44 valence electrons. The remaining 26 are then located in the seven bonding molecular orbitals $d_z^2(a_1')^2$ $[d_{xz}](e_1')^4$ $[d_{x^2-y^2}](a_1')^2$ $[d_{yz}](a_2'')^2$ $[d_{xy}](e'')^4$ and the lower lying six of the eight antibonding ones $[d_{xy}](a_1''^*)^2$ $[d_{yz}](e''^*)^4$ $[d_{x^2-y^2}](e'^*)^4$ $[d_{xz}](a_2'^*)^2$ $[d_z^2](e'^*)^0$.

Using simple overlap arguments based on the shape of the above-indicated molecular orbitals, the highest occupied molecular orbital is suggested to be the $a_2'^*$, antibonding combination of d orbitals in the xz plane. Let us consider that to formulate this hypothesis only the molecular symmetry that means the symmetry operators has been used. The use of Hamiltonian operator, really very difficult, has been substituted by the hypothesis that the electronic structure parallels the molecular geometry. It is now mandatory to collect a number of experimental data to demonstrate this hypothesis.

5.2 Discussion of the Case

5.2.1 X-Ray Diffraction Studies

The analysis shows a significant shortening of the Pt–Pt distance going from $[Fe_3\,Pt_3\,(CO)_{15}]^{2-}$ (2.750 Å) to $[Fe_3\,Pt_3\,(CO)_{15}]^{1-}$ (2.656 Å). This suggests that the insertion of one additional electron in $[Fe_3\,Pt_3\,(CO)_{15}]^{2-}$ induces shortening of the Pt–Pt bond. As a consequence, the additional electron should be located in an antibonding orbital, lying along the Pt–Pt bond that is the $a_2'^*$ antibonding combination of d orbitals in the xz plane.

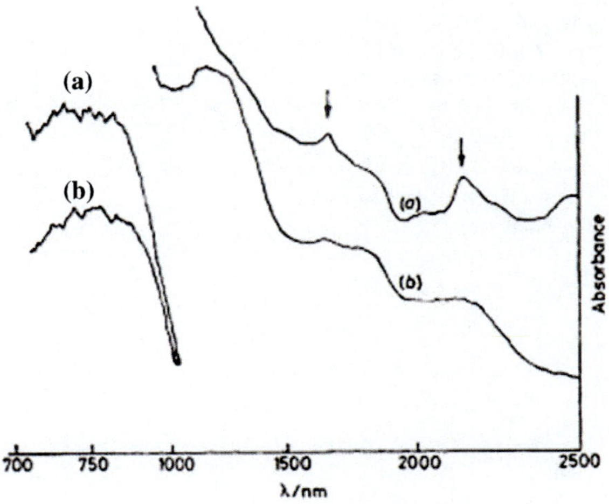

Fig. 5.2 Electronic diffuse reflectance spectra of **a** $[Fe_3\ Pt_3\ (CO)_{15}]^{2-}$, **b** $[Fe_3\ Pt_3\ (CO)_{15}]^{-}$ [3]

5.2.2 *Diffuse Reflectance Spectra*

The spectra of $[Fe_3\ Pt_3\ (CO)_{15}]^{n-}$ (n = 1, 2) (Fig. 5.2) are only available for qualitative purposes. An absorption at 1250 nm distinguishes the spectrum of the monoanion from that of the dianion, while both have strong absorption at 700 nm. Either the energy or the intensity of these bands indicates attribution to **d-d** electronic transitions between the molecular orbitals deriving from the combination of **5d** Pt orbitals. It has been suggested that the band at 1250 nm is due to the electron transfer to the $a_2'^*$ antibonding orbital and it is active only in case of half occupation of the same orbital.

5.2.3 *Electron Spin Resonance (ESR) Spectra*

These are recorded only for the paramagnetic unit $[Fe_3\ Pt_3\ (CO)_{15}]^{-}$. The spectrum recorded at −150 °C (Fig. 5.3) shows anisotropic resonance lines separated into two different intensity groups of well-detectable seven-line splitting in the higher field group.

It may be noted that the spectrum indicates an electronic situation greatly affected by the high value of the platinum spin–orbit coupling perturbation and that the g anisotropy $\Delta g = 0.51$ is comparable to that of the monomeric Pt compounds. This suggests that the unpaired electron is mainly located on Pt centers. The hyperfine splitting with **A** constant is due to the interaction of the unpaired electron with the ^{195}Pt nucleus, I = 1/2, which occurs naturally with 33.8% abundance,

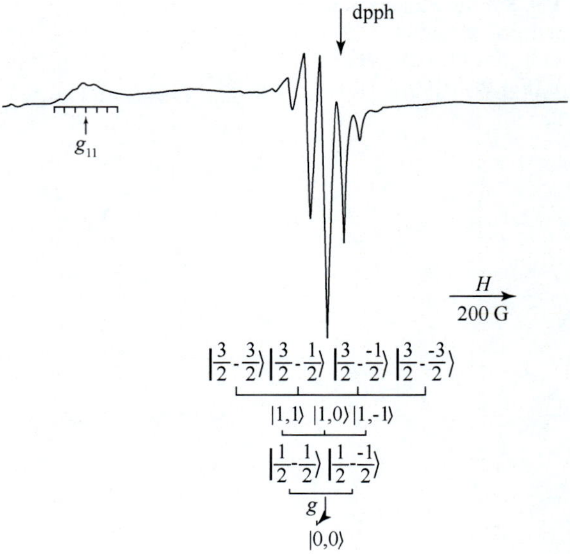

Fig. 5.3 ESR spectrum of $[Fe_3\ Pt_3\ (CO)_{15}]^-$ in frozen solution. Transitions have been assigned by their I I M_I > values; $g_{\parallel} = 2.57$, $g_{\perp} = 2.03$; $A_{\parallel}$ (G) = 60, $A_{\perp}$ (G) = 100 (G is the Gauss unit) [3]

while the remaining Pt isotopes have I = 0. The ESR spectrum results in a superimposed singlet, doublet, triplet, and quartet having, respectively, 0, 1, 2, and 3 ^{195}Pt nuclei in the central Pt_3 triangle.

5.3 Chemical Message

From the discussion of the data given by structural and spectromagnetic analyses, it is conceivable that the paramagnetic $[Fe_3\ Pt_3\ (CO)_{15}]^-$, though containing one unpaired electron and in principle very reactive, should be stable, due to both the delocalization of the unpaired electron on the Pt_3 triangle and the steric hindrance of the ligands which oppose the pairing.

In fact, when considering the condensation of two $[Fe_3\ Pt_3\ (CO)_{15}]^-$ cluster units into the $[Fe_4\ Pt_6\ (CO)_{22}]^{2-}$, the absence of one of the three Fe-centered moieties at one triangle corner allows pairing of the two Pt_3 units (Fig. 5.4).

This behavior could also suggest that the nuclearity (number of bonded metal centers) of transition metal clusters depends on the coupling of paramagnetic fragments with lower nuclearity. This is the message. Literature [4] reports other examples of growing in cluster nuclearity associated with the presence of unpaired electrons. Thus, it seems that the fragmentation of a pristine cluster and the formation of units having paramagnetic centers and controlled steric hindrance is a way to proceed from the molecule to the particle.

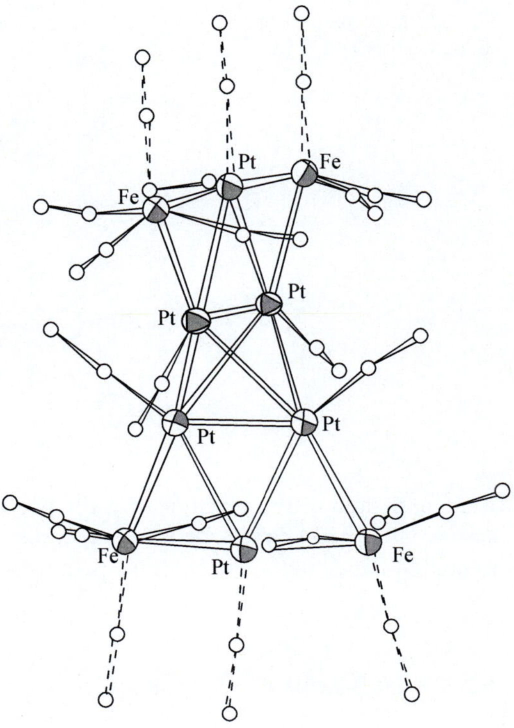

Fig. 5.4 Molecular structure of the diamagnetic $[Fe_4Pt_6(CO)_{22}]^{2-}$

5.4 The Case of Adducts of NN′Ethylenebis (acetylacetoneiminato)Cobalt (II)

[NN′Ethylenebis(acetylacetoneiminate)cobalt(II)] (CoAcacen) and [NN′ethylenebis (salycilideneiminate)cobalt(II)] (CoSalen) have been studied for a long time, due to their ability to coordinate molecular oxygen in a reversible way [5–8]. This property suggested that the complexes could be model of the natural oxygen carriers, e.g., hemoglobin. Due to their relatively simple structure, these cobalt complexes allow to study the electronic modifications connected to the O_2 absorption. For this reason, it was carried out an extensive study on the pristine CoAcacen and CoSalen and on their adducts with base molecules, typically pyridine (very reactive adduct toward oxygen chemisorption). The study, performed by structural, electronic, and spectromagnetic investigations, suggested the electronic configurations of the Co (II) active to interact with molecular oxygen, at the same time transferring one electron to O_2 and giving rise to a superoxide derivative.

Figure 5.5 reports the structure of CoAcacen, planar molecule belonging to C_{2v} symmetry. The question is why this reacts so slowly with O_2 forming [Co(III) Acacen O_2^-] while, if the coordination of a base molecule takes place as in [Co(II)

Fig. 5.5 Molecular scheme of *M(II)Acacen* M = Co

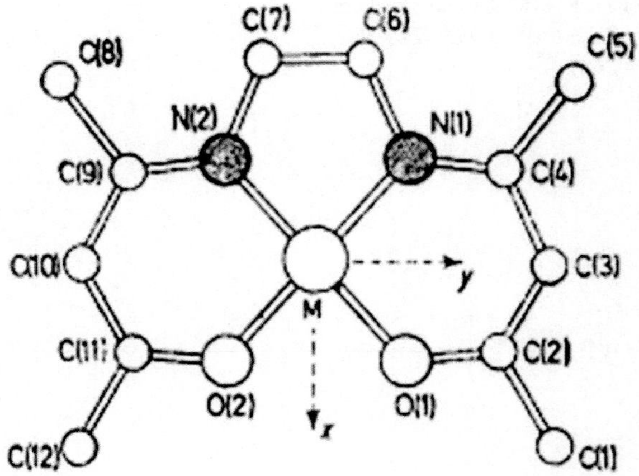

Acacen pyridine], the reaction to form [Co(III)Acacen PyO_2^-] is very fast? To answer, it is necessary to perform experiments that reveal which electronic configuration the cobalt assumes in the pristine and in the base adduct.

5.5 Discussion of the Case

The electron spin resonance investigation was performed for CoAcacen on a single crystal sample, diluted with the NiAcacen diamagnetic host. In order to simplify, the calculations were performed by a three-hole model, instead of the more complicated seven-electron configuration.

The following expressions report the g magnetic tensor components as a function of the distance of the excited states from the ground state, respectively, from $(xz)^2(xy)^2(yz)(z^2)^2$ and $(xz)^2(xy)^2(yz)^2(z^2)$ assumed as ground states. In parentheses are indicated the d orbitals, xz, xy, yz, z^2. The experimental g tensors were reproduced for the two different ground state hypotheses, and the calculated values are in parentheses under the experimental ones.

c and **b** coefficients indicate the weight of the different excited states within the ground state and are derived from the perturbation of the ground state by the spin–orbit operator.

Ground configuration $(xz)^2(xy)^2(yz)(z^2)^2$

$$g_{xz} = 2c_5^2 + 2c_2^2 - 2c_3^2 - 2c_1^2 - 2c_4^2 - 4c_2c_5$$

$$g_{xx} = 2c_5^2 - 2c_2^2 - 2c_3^2 + 2c_1^2 + 2c_4^2 - 4\sqrt{3}c_1c_5 - 4c_2c_3 + 4c_4c_5$$

$$g_{yy} = 2c_5^2 - 2c_2^2 + 2c_3^2 - 2c_1^2 - 2c_4^2 - 4c_3c_5 + 4\sqrt{3}c_1c_2$$

g_{zz}	g_{xx}	g_{yy}
2.00	3.26	1.88
(1.87)	(3.24)	(2.05)

c_1	c_2	c_3	c_4	c_5
−0.2064	−0.0359	−0.0664	−0.0035	0.9556
$E_{yz} - E_{z^2}$	$E_{yz} - E_{zz}$	$E_{yz} - E_{xy}$	$E_{yz} - E_{x^2-y^2}$	
(*ca.* 1700)	8000	5000	(57.000)	
	(*ca.* 6000)	(3000)		

Ground configuration (xz)^2(xy)^2(yz)^2(z^2)

$$g_{zz} = 2b_3^2 - 2b_2^2 - 2b_1^2 + 4Kb_1b_2$$
$$g_{xx} = 2b_1^2 - 2b_2^2 + 2b_3^2 - 4\sqrt{3}Kb_1b_3$$
$$g_{yy} = 2b_3^2 + 2b_2^2 - 2b_1^2 - 4\sqrt{3}Kb_2b_3$$

g_{zz}	g_{xx}	g_{yy}
2.00	3.26	1.88
(1.89)	(3.21)	(2.12)
b_1	b_2	b_3
−0.1803	−0.0380	0.9829
$E_{z^2} - E_{yz}$	$E_{z^2} - E_{xz}$	
(1900)	8000	
	(*ca.* 9000)	

The research pathway consists in calculating the g component values by a second-order approximation, using the g expressions above reported and the crystal field energy differences derived from the electronic absorption spectra of planar CoAcacen and pentacoordinated CoAcacen pyridine (Fig. 5.6, the energies are in cm–1) and then in comparing the calculated g values with the experimental ones (reported in parentheses).

It can be observed that whatever is the ground state configuration, the experimental values of g_{zz} and g_{yy} cannot be satisfactorily reproduced. Moreover, the experimental crystal field energies fit with both the ground state configurations. For these reasons, it is not possible to unambiguously assign the ground state configuration, and a large mixing between the states *(xz)^2(xy)^2(yz)(z^2)^2* and *(xz)^2(xy)^2(yz)^2(z^2)* seems to occur. The relative distance is less than 2000 cm^{-1}.

In any case, the unpaired electron lies out of the molecular plane and can be easily transferred to the π^* molecular orbitals of the coordinated oxygen. Even a feeble axial perturbation along the z-axis could stabilize the *(xz)^2(xy)^2(yz)^2(z^2)* ground configuration, which becomes really very suitable to react with oxygen, due to the location of the unpaired electron in the d_z^2 orbital. This happens in [Co(II) Acacen py] where the absorption at about 8000 cm^{-1} shifts to about 12,000 (Fig. 5.6).

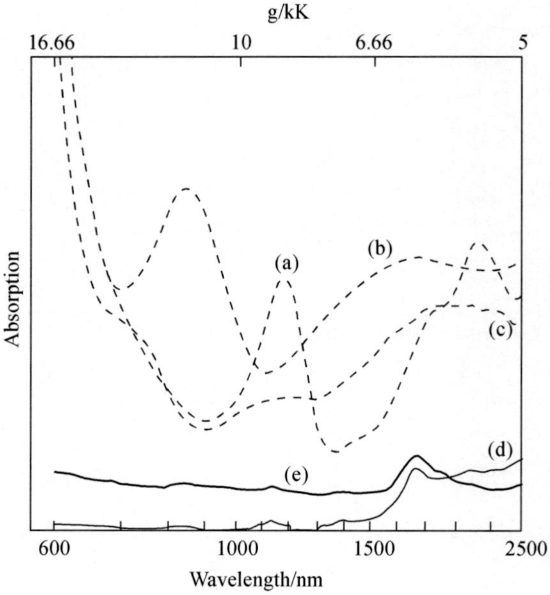

Fig. 5.6 Diffuse reflectance spectra reported in nm and in kK ($cm^{-1} \times 1000$) of **a** planar CoAcacen and **b** pentacoordinated CoAcacen py [5–8]

The change of ground state configuration of the cobalt(II) is very evident, comparing the spectrum (a) with (b), and indicates that the coordination of pyridine increases the weight of the $(xz)^2(xy)^2(yz)^2(z^2)$ configuration. Consequently, the electron transfer from Co(II) to O_2 becomes easier.

5.6 Chemical Message

From the above discussion, it appears that the responsibility of the easy electron transfer from the Co(II) to oxygen lies in the location of the unpaired electron along the z-axis. However, also the $(xz)^2(xy)^2(yz)(z^2)^2$ configuration assists this interaction through overlap of the $d_{xz,yz}$ with the π^* orbitals. When oxygen is removed, the unpaired electron reversibly returns onto the cobalt.

5.7 The Case of Phthalocyaninato Co(II)

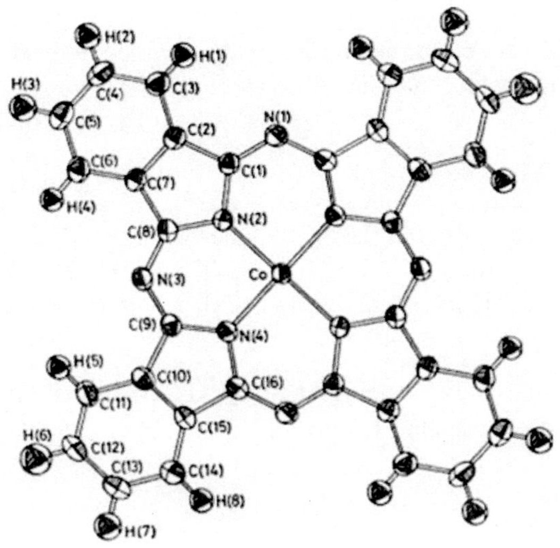

Structure of Phtalocyaninato Co(II)

This Co(II) coordination compound has been widely studied due to its structural similarity with the porphyrin complexes, which in turn are responsible for fundamental interaction with oxygen in human life [9, 10]. As it concerns the reaction with oxygen, the most fascinating aspect stays in the electronic distribution between O_2 and Co(II): in fact, it can be modeled as Co(II)-O_2, with the unpaired electron on the cobalt, or as Co(III)-O_2^- with the electron on the oxygen, to stabilize the superoxide anion. The decision between the two options is an old scientific problem, and it begun since from the pristine work of L. Malatesta on cobalt amino derivatives. The scientist suggested the following possibilities

$$[\mathrm{Co(III)\text{-}O_2\text{-}Co(IV)}]^{5+} \quad [\mathrm{Co(III)\text{-}O_2\text{-}Co(III)}]^{5+}$$

while he could not demonstrate on where the unpaired electron was located.

In general, this query is very important because porphyrin and porphyrin-like compounds are able to reversibly adsorb oxygen and the electron transfer process is a knowledge requirement.

A proper model molecule to be studied is the [Phthalocyaninato Co(II)] (CoPc). Its structural similarity with the heme proteins suggested an easy model role for it; moreover, its ability to reversibly bind molecular oxygen and to modulate this reactivity, as a function of a coordinated base molecule (e.g., pyridine), strongly prompted to study its electronic and magnetic properties.

5.8 Discussion of the Case

5.8.1 Electron Spin Resonance (ESR) Spectra

The ESR spectra were performed on the pristine compound and on its adduct with base molecules (4-methyl pyridine, 4Mepy) (Fig. 5.7).

The g values are:

	$g_{xx} = g_{yy}$	g_{zz}	$b_1 = b_2$	b_3
$[CoPc]$	2.94	1.89	−0.1531	0.9763
$[CoPc(4Mepy)]$	2.32	2.005	−0.0475	0.9977
$[CoPc(4Mepy)_2]$	2.25	2.015	−0.0368	0.9986

b_1 b_2 b_3 are, respectively, the weight of d_{xz} d_{yz} d_z^2 orbitals in the ground state configuration and are calculated from

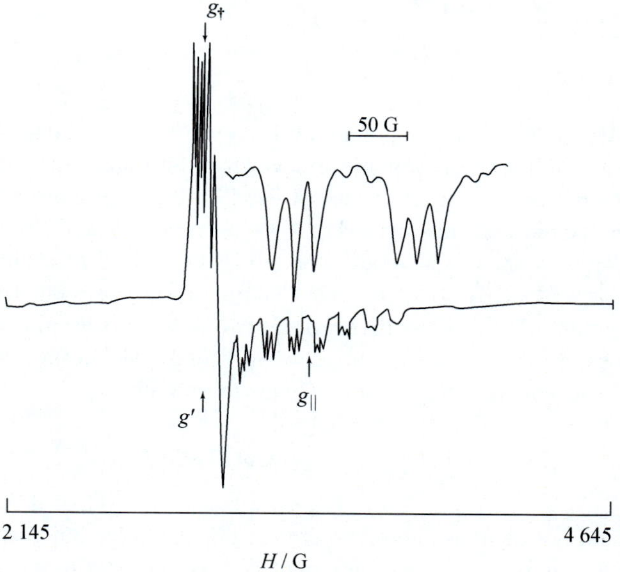

Fig. 5.7 ESR spectrum of [CoPc(4-methyl pyridine)]. $\mathbf{g}_{\|} = \mathbf{g}_{zz}$ $\mathbf{g}_{\dagger} = \mathbf{g}_{xx,yy}$ $\mathbf{g}'$ is due to impurity. In the $\mathbf{g}_{\|}$ region it is well evident the hyperfine interaction with nitrogen nucleus of 4-methyl pyridine [9, 10]

$$g_{zz} = 2b_3^2 - 2b_2^2 - 2b_1^2 + 4Kb_1b_2$$
$$g_{xx} = 2b_1^2 - 2b_2^2 + 2b_3^2 - 4\sqrt{3}Kb_1b_3$$
$$g_{yy} = 2b_3^2 + 2b_2^2 - 2b_1^2 - 4\sqrt{3}Kb_2b_3$$
$$K = 1$$

The unpaired electron results to be located in the d_z^2 orbital and the contribution of this orbital increases with the base coordination, while those of d_{xz} and d_{yz} decrease. The g_{zz} value increases as the axial coordination becomes stronger. This electronic asset is responsible for stabilizing $[CoPc(4Mepy)\ O_2]^-$ where the unpaired electron is located on O_2. The experimental proof comes from the ESR spectrum, reported in Fig. 5.8 which shows $g_{xx} = g_{yy} = 2.01$ and $g_{zz} = 2.00$, very different from the magnetic tensor values of the CoPc adducts with base molecules. The values are diagnostic of an unpaired electron strongly located onto the superoxide radical anion, and the ESR investigation strongly answers the doubt expressed by Malatesta.

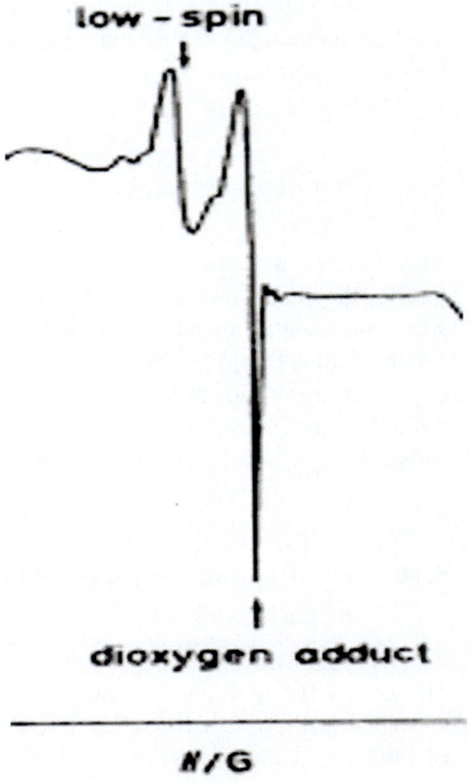

Fig. 5.8 ESR spectrum of $[CoPc(4Mepy)O_2]^-$ in the center of the field range. The resonances are centered at about 3000 G. At lower field the resonances of [CoPc (4Mepy)]

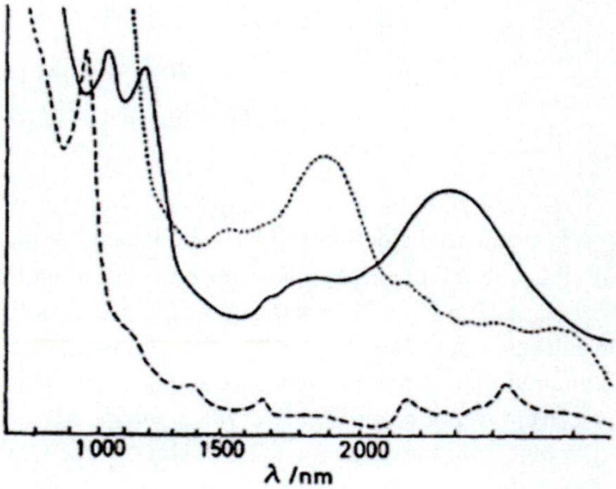

Fig. 5.9 Diffuse reflectance spectra of __βCoPc …αCoPc - - -CoPc 4Mepy (α and β are polymorphous forms where the α one contains intermolecular axial interaction between the two Pc rings) [9, 10]

5.8.2 *Diffuse Reflectance Spectra*

The diffuse reflectance spectra, taken in the near-infrared region, are in agreement with the absorption calculated from the ESR spectra.

The calculated energy (from the g values) of the $d_{xz,yz} \rightarrow d_z^2$ increases with the axial coordination in connection with the decrease of $g_{xx,yy}$ in agreement with the resonance spectrum in Fig. 5.9.

5.9 Chemical Message

At a variance with the CoAcacen, CoPc displays a ground state well suitable to transfer the unpaired electron to the oxygen molecules, provided the coordination site is not occupied by other previously coordinated base molecules (e.g., in the bis adduct $[CoPc(4Mepy)_2]$). This confirms in CoPc the same mechanism of the oxygen chemisorption as in CoAcacen. The mechanism could be suggested also for the natural porphyrin complexes, responsible for the modification of the chemisorbed oxygen, which accepts and reversibly returns the unpaired electron.

5.10 The Case of Oxygen Reduction by CoPc Supported on Inorganic Oxide

The progresses in heterogeneous catalysis with porphyrin-like complexes have prompted a direct spectroscopic investigation on the catalyst surface [11]. CoPc supported on ϒ Al_2O_3 has been investigated in order to understand the role of these heterogeneous catalysts in different types of oxidative processes, e.g., hydrocarbon oxidation, and to investigate the relations between surface chemisorbed species and catalytic properties. It is well known that CoPc in solution gives superoxo derivatives where the unpaired electron of Co(II) d^7 is transferred to the O_2 molecule. The possibility of an analogous electron transfer in the solid state was the basis for the investigation.

5.11 Discussion of the Case

The solid catalyst [CoPc.ϒAl_2O_3] contacted with increasing pressures of oxygen showed the resonance lines in Fig. 5.10, whose intensities increase with the gas pressure. The g values $g_{\parallel} = 2.034$ and $g_{\dagger} = 2.003$ are in agreement with those found when O_2^- is chemisorbed on a triply charged cation, that is, Al^{3+}-O_2^-. It is hypothesized that an electron transfer from the complex CoPc to the oxygen originates the O_2^- anion, which then fixes onto Al^{3+} centers. If Al_2O_3 is substituted by SiO_2, the superoxide anion is not visible, suggesting that the ability of an oxide to fix negative charges is a prerequisite.

5.12 Chemical Message

The activation of CoPc to react with oxygen also in the solid [CoPc]. ϒAl_2O_3 has the same origin as in the liquid phase. The ease of absorption is probably still here related to an increased weight of the $(x^2 - y^2)^2(xz, yz)^4(z^2)$ configuration in the

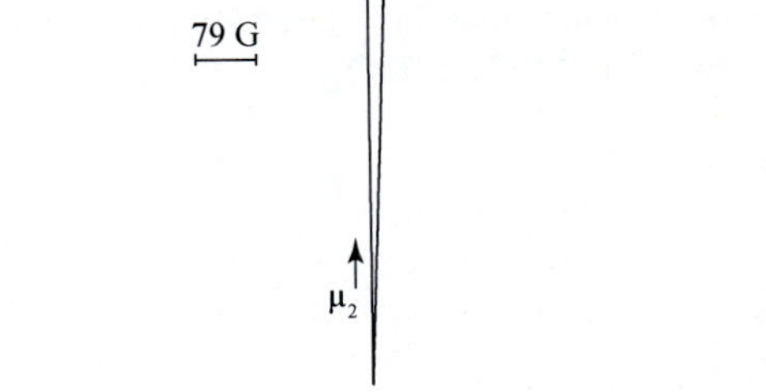

Fig. 5.10 ESR spectrum of CoPc Al_2O_3, contacted with O_2 40 atm. [11]

ground state of cobalt. This is due to σ donation of the oxide anion O^{2-} of ΥAl_2O_3 to the cobalt, along the z-axis of the complex.

5.13 The Case of Metal-Dispersed Inorganic Oxides

Since from the origin of the studies in heterogeneous catalysis, it appeared that the metal centers, mainly transition metal centers, are responsible for the catalytic activity [12]. This is due to an ease electron transfer assisted by the metals from the reactant to the substrate. Nevertheless, the electronic structure of the metal centers remained difficult to be known, mainly as the metals are present in high concentration and poorly diluted state, while the spectroscopic methods better operate on dispersed systems. A sophisticated technique, based on preparing dispersed metal samples by thermal decomposition of carbonyl metal clusters, allowed to prepare dispersed metals in inorganic oxides. These samples are suitable to be investigated by spectroscopic methods as pristine or after reaction with the substrate molecules. We report the investigation of $[Co_4(CO)_{12}]$ and $[Rh_4(CO)_{12}]$ carbonyl clusters dispersed on ΥAl_2O_3 and ZrO_2 supports, then thermally treated to give dispersed metal systems.

5.14 Discussion of the Case

5.14.1 Electron Spin Resonance of Decarbonylated Systems

Co. $\boldsymbol{\Upsilon Al_2O_3}$ under argon atmosphere gave low-intensity signals, not worth to be considered. ***Rh.*** $\boldsymbol{\Upsilon Al_2O_3}$ in the same conditions instead gave strong asymmetrical lines at $g_{\parallel} = 2.21$ and $g_{\dagger} = 2.11$, which suggest location of the unpaired electron on Rh centers. The hyperfine coupling interaction with Rh nucleus (I = ½) is also visible (Fig. 5.11).

The values of the g tensor components with $g_{\parallel} > g_{\dagger} > 2$ indicate a Rh(II) d^7 with the unpaired electron in the $d_{x^2-y^2}$ orbital, in agreement with an oxidative interaction between Rh and ΥAl_2O_3.

If the same experiments are performed on zirconia-dispersed metals, no paramagnetic species are detected. This suggests that the metal centers interact with the oxide support and the interaction is as more oxidative as stronger the acidity of the support, $Al_2O_3 > ZrO_2$.

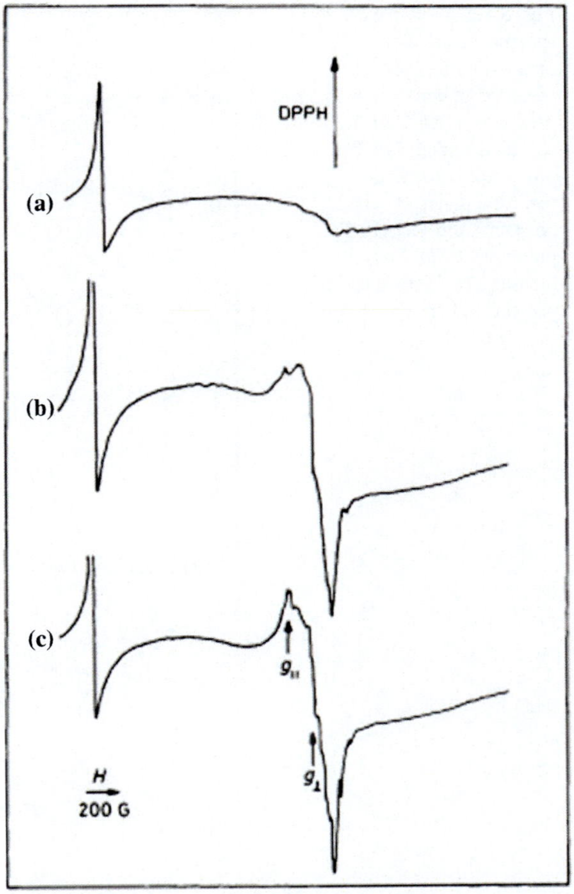

Fig. 5.11 X band ESR spectra of **a** ΥAl_2O_3 and **b** Rh.ΥAl_2O_3 from pyrolysis of $[Rh_4(CO)_{12}]$ ΥAl_2O_3 at 150 °C, **c** Rh.ΥAl_2O_3 from pyrolysis of $[Rh_4(CO)_{12}]$ ΥAl_2O_3 at 250 °C [12]

5.14.2 Interaction with Oxygen

***Co.*ΥAl_2O_3** was contacted with O_2 at room temperature for 5 min. The X band ESR spectrum of the sample is reported in Fig. 5.12, line e.

The resonance lines in (e) $g_{\|} = 2.034$ and $g_{\dagger} = 2.00$ indicate the presence of the Al^{3+}-O_2^- species.

***Rh* ΥAl_2O_3** treated by O_2 shows asymmetric resonance lines, $g_1 = 2.09$, $g_2 = 2.03$, $g_3 = 2.00$; these suggest that the unpaired electron occupies a O_2 π^* orbital with a small contribution of a d_π Rh orbital (line c).

***Rh* ZrO_2** reveals the presence of the same species as on ***Rh* ΥAl_2O_3** together with the species Zr^{4+}-O_2^-.

***Co* ZrO_2** did not show resonances.

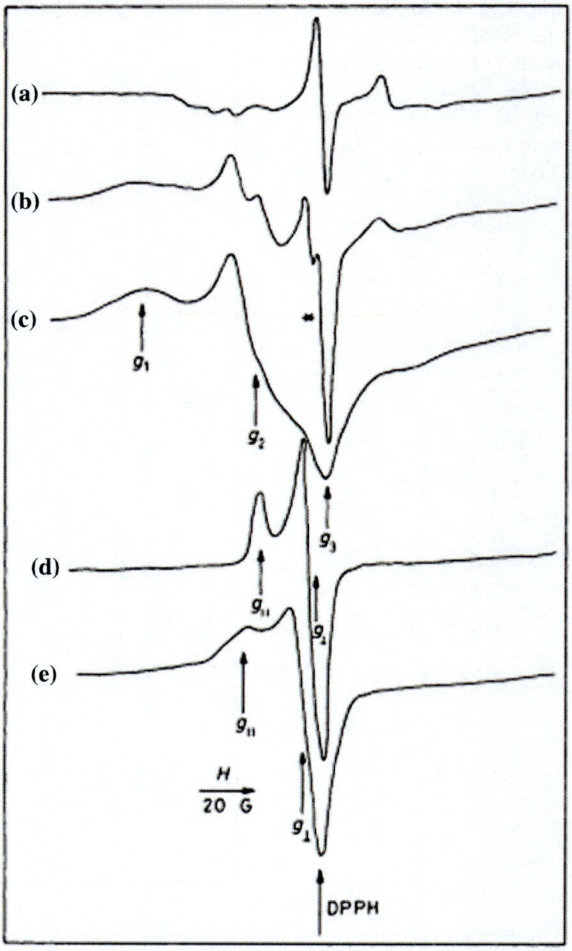

Fig. 5.12 X band ESR spectra recorded at −150 °C of **a** $\boldsymbol{ZrO_2}$, **b** ***Rh*** $\boldsymbol{ZrO_2}$ recorded under O_2 (10 Pa pressure), **c** ***Rh*** $\boldsymbol{YAl_2O_3}$ recorded under O_2 (10 Pa pressure), **d** ***Rh*** $\boldsymbol{ZrO_2}$ recorded under O_2 (10 Pa pressure) and CO (10 Pa pressure), **e** ***Co*** $\boldsymbol{YAl_2O_3}$ recorded under O_2 (10 Pa pressure) [12]

5.14.3 Interaction with Carbon Monoxide

Among ***Co*** $\boldsymbol{YAl_2O_3}$ ***Rh*** $\boldsymbol{YAl_2O_3}$ ***Rh***$\boldsymbol{ZrO_2}$, only ***Rh*** $\boldsymbol{YAl_2O_3}$ shows a Rh-CO adduct (Fig. 5.13).

The spectrum in Fig. 5.13, due to the wideness of lines, indicates the unpaired electron on the Rh centers; nevertheless, the small displacement from the free electron value and the absence of magnetic anisotropy suggest that the interaction with CO removes part of the unpaired electron from the metal center.

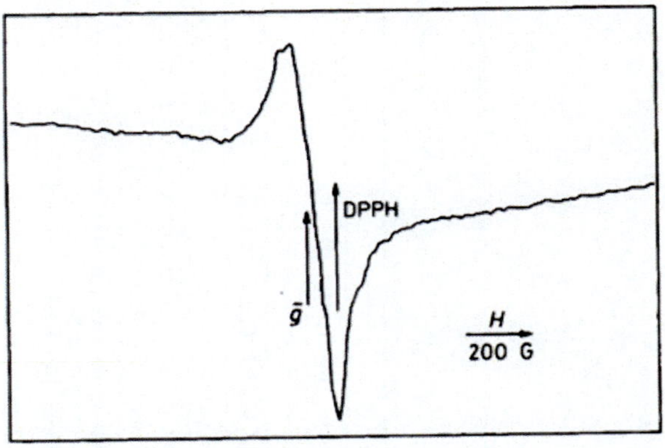

Fig. 5.13 X band ESR spectrum of ***Rh YAl₂O₃*** under CO (10 Pa atmosphere) [12]

5.14.4 *Interaction with Carbon Monoxide and Oxygen*

Table 5.1 suggests that the formation of paramagnetic species by decarbonylation of supported clusters depends both on the metal and on the oxide. YAl_2O_3 has higher acidic character and stabilizes oxidized Rh species. Moreover, the interaction between Rh and oxide is more efficient due to the better overlap between Al^{3+} p orbitals and the very expanded Rh d_π orbitals. The results concerning the chemisorption of O_2 and CO show that the superoxide is fixed both and on Rh and in Al^{3+} or Zr^{4+} centers due to the higher strength of the ionic interaction.

5.15 Chemical Message

One of the most important problems that justify the investigation of supported metal systems used as catalysts is concerning the knowledge of the metal oxidation state. There seems to be evident that the selectivity of such systems depends on metal centers having positive oxidation state. Conventional catalysts, obtained by reduction of supported metal ions, show in general few oxidized centers, whose amount is limited to centers interacting with the support. Thus, the formation of large metal particles after reduction hinders the interaction and lowers the reactivity of the catalytic sites. It seems reasonable that when the metal precursor is a pyrolyzed cluster, being the metal on the surface a polynuclear system, the positive charge generated by interaction with the support can be distributed among all singly interacting metal centers. This increases the availability of active catalytic centers.

Samples *Rh* YAl_2O_3 and $RhZrO_2$ have been tested for the CO + H_2 reaction and show that: YAl_2O_3-supported samples produce methane and higher hydrocarbon, and ZrO_2 produces oxygenated products. The ESR results show that the interaction of π^* acceptor molecules such as CO and O_2 with Rh is stronger on Al_2O_3 than on

Table 5.1 ESR data for all the metal supported

Sample	O_2 pressure (Pa)	CO pressure (Pa)	g_{zz}	g_{zx} or yy	g_{yx} or xx	No. of paramagnetic centers/spin (g^{-1})	Paramagnetic species[e]	Notes
$Co_4(CO)_{12}$/ γ-Al_2O_3	0	0	2.00	2.00	2.00	0.28×10^{14}	Radical	–
	10	0	2.034	2.00	2.00	0.56×10^{12}	$Al^{3+} - O_2^-$	Stable at 180 °C and 10^{-2} Pa
	0	10	–	–	–	–	–	–
	10	10	2.034	2.00	2.00	1.12×10^{16}	$Al^{3+} - O_2^-$	Stable at 180 °C and 10^{-2} Pa
$Rh_4(CO)_{12}$/ γ-Al_2O_3	0	0	2.21	2.11	2.11	0.56×10^{16}	Rh^{11}	Multinuclear species
	10	0	2.09	2.03	2.00	0.46×10^{16}	$Rh^{[111]} - O_2^-$	Thermally removed at 250 °C, 10^{-2} Pa with irreversible sample modification
	0	10	2.034	2.034	2.034	0.28×10^{14}	Rh^{11} – CO	Thermally removed at 100 °C, 10^{-2} Pa
	10	10	{ 2.034 2.034	2.00 2.034	2.00 2.034	0.42×10^{12} 1.38×10^{12}	$Al^{3+} - O_2^-$ Rh^{111} – CO }	Stable at 100 °C and 10^{-2} Pa
$Rh_4(CO)_{12}$/ ZrO_2	0	0	–	–	–	–	–	–
	10	0	{ 2.09 2.03	2.03 2.00	2.00 2.00 }	0.16×10^{16}	{ $Rh^{[1]} - O_2^-$ $Zr^{4+} - O_2^-$ }	Removed by vacuum treatment at 10^{-3} Pa, reversible process
	0	10	–	–	–	–	–	–
	10	10	2.03	2.00	2.00	0.12×10^{16}	$Zr^{4+} - O_2^-$	On vacuum treatment (10^{-3} Pa) $Rh^{[1]} - O_2^-$ is informed as well as $Zr^{4+} - O_2^-$

ZrO_2. As a consequence on Al_2O_3, the stronger metal–CO interaction weakens the C–O bond and leads to hydrocarbon formation; on ZrO_2, the weaker interaction cannot lead to dissociation of the C–O bond and oxygenated products are formed.

5.16 The Case of Phosphate Glasses Containing Ruthenium and Titanium Ions

Oxide glasses containing transition metals have attracted interest for a long time due to their behavior as disordered semiconductors [13]. Their semiconductive properties are interpreted as being due to the presence of mixed valence transition metal ions. According to Austin and Mott, the electronic behavior of these systems could be well described in terms of thermally assisted hopping of electrons; however, this model is able to give only qualitative description of the conduction processes. It seems mandatory the detection of the different oxidation states of the transition metal ions dispersed in the glasses.

5.17 Discussion of the Case

Glasses containing TiO_2 as well as RuO_2 were obtained from mixtures of Na_2TiO_3, sodium metaphosphate, and orthophosphoric acid melted at 1260 °C. Reduced titanium and ruthenium centers, present near the major amount of oxidized ones, have been detected by electron spin resonance investigation, and the g tensor values are in accordance with Ru^{3+} and Ti^{3+} species (Fig. 5.14).

The trends in the amount of reduced Ti and Ru centers (Ti^{3+}, Ru^{3+}) allow to acknowledge the distribution of the transition metal ions in samples having different concentrations and to design the samples with the desired electric characteristics.

5.18 Chemical Message

The presence of two different oxidation states for the doping transition metal ions is a prerequisite to control the electric conductivity, based on the Austin and Mott model. The reported investigation allows to detect the magnetic dilution of the paramagnetic centers and to predict the conditions suitable for the hopping mechanism.

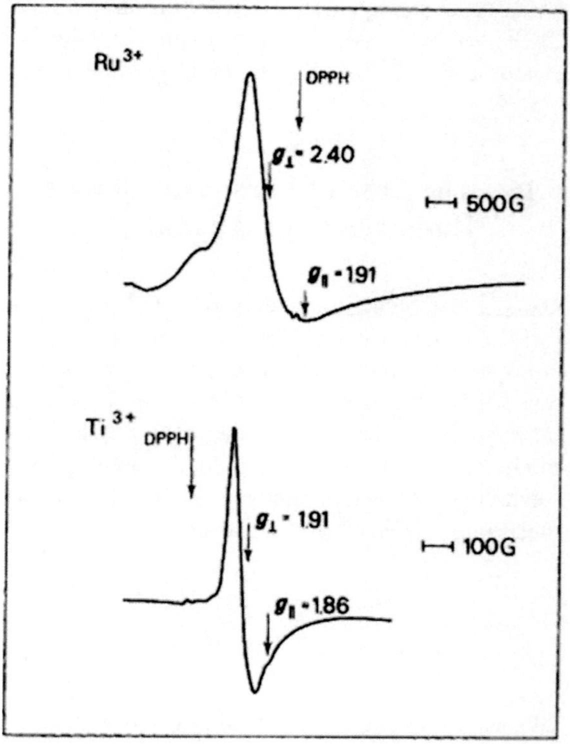

Fig. 5.14 ESR spectra of RuO_2 and TiO_2 containing phosphate glasses [13]

Table 5.2 ESR behavior of differently doped glasses

$[TiO_2]$ (mol%)	$[RuO_2]$ (mol%)	$g_{\parallel}$	$g_{\perp}$	Paramagnetic species	10^8 $[M^{3+}]$ (at%)	$10^8 \frac{[M^{3+}]}{[M^{4+}]}$
2.47	–	1.86	1.91	Ti^{3+}	11	0.5
3.04	–	1.86	1.91	Ti^{3+}	19	6
3.00	–	1.86	1.91	Ti^{3+}	5	2
3.45	–	1.86	1.91	Ti^{3+}	33	9
4.10	–	1.86	1.91	Ti^{3+}	50	13
5.00	–	1.86	1.91	Ti^{3+}	24	5
7.80	–	1.86	1.91	Ti^{3+}	96	12
12.00	–	1.86	1.91	Ti^{3+}	445	37
4.80	0.10	1.91	2.40	Ru^{3+}[a]	958	9580
5.00	0.20	1.91	2.40	Ru^{3+}[a]	1240	6050
4.98	0.19	1.91	2.40	Ru^{3+}[a]	1130	5947
5.00	0.19	1.91	2.40	Ru^{3+}[a]	1050	5526

[a] Ti^{3+} species have not been detected

Table 5.3 ESR behavior of differently doped Ru-based glasses

Sample no.	$[RuO_2]$ (mol%)	g_l	$g_\perp$	Paramagnetic species	10^5 $[Ru^{3+}]$ (at%)	$10^5 \frac{[Ru^{3+}]}{[Ru^{4+}]}$
72	0.10	1.91	2.40	Ru^{3+}	2.4	24
89	0.26	1.91	2.40	Ru^{3+}	5.2	20
75	0.43	1.91	2.40	Ru^{3+}	7.7	19
50	0.78	1.91	2.40	Ru^{3+}	8.9	13
39	1.19	1.91	2.40	Ru^{3+}	16.0	13
67	1.87	1.91	2.40	Ru^{3+}	31.0	16
46	2.40	1.91	2.40	Ru^{3+}	35.0	14
71	2.94	1.91	2.40	Ru^{3+}	48.0	16
60	3.50	1.91	2.40	Ru^{3+}	71.0	20

5.19 The Case of Superoxide Dismutase Enzyme, Interacting with Antitumor Drugs

Several investigations concerning with the definition of the activity of the copper–zinc superoxide dismutase (Cu-Zn SOD) have been carried out, and the emerging consensus is that the enzyme has usually, but not always, a lower activity in tumor cells, with respect to normal cells [14]. Although the implication for carcinogenesis and cancer chemotherapy of this activity loss is still unknown, this experimental finding may be of importance in understanding the mode of action of antitumor agents. There is a large class of cancer chemotherapeutic compounds that require as a step in their reaction mechanism the production of superoxide O_2^- and/or other oxygen-centered radicals, e.g., $HO^{\cdot}$, $HO_2^{\cdot}$, etc. These oxygen species are thought to be responsible for the DNA damage that results in the cell's destruction. Based on the observation of a lowered SOD activity in cancer cells, one could anticipate that an increased production of superoxide (e.g., induced by antitumor drugs) should result in a preferential destruction of tumor cells. In addition, an antitumor drug could further reduce the SOD activity by chelation with the enzyme active metal site, e.g., Cu^{2+}. Based on the works which considered the interaction between Cu^{2+} and anthracyclines, the investigation was extended to the system Cu-Zn SOD—anthracycline with the aim of assessing the accessibility of copper centers to the drug molecules.

5.20 Discussion of the Case

Figure 5.15 reports two series of spectra concerning the interaction of Cu-Zn SOD with doxorubicin, at different concentrations and for different r = *drug/enzyme* molar ratios. The pattern is representative of the Cu-Zn SOD, $g_{\parallel} = 2.272$ and

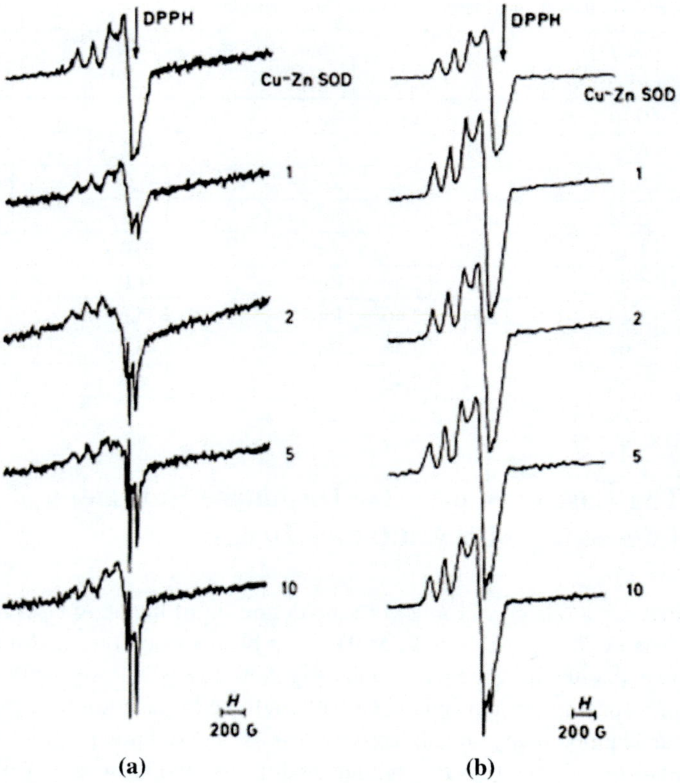

Fig. 5.15 ESR spectra of Cu-Zn SOD doxorubicin at pH 6.5 for **a** Cu-Zn SOD 0.125 mmol dm^{-3} and **b** Cu-Zn SOD 0.675 mmol dm^{-3} both for the drug/enzyme molar ratios right-hand-indicated [14]

$g_{\dagger} = 2.078$ $A_{\parallel} = 130$ G, until r = 2 or r = 5 for the two concentrations of Cu-Zn SOD. Then (r = 10) it appears a new species which can be associated with Cu doxorubicin complex (Fig. 5.16). This means that the antitumor drug is able to extract copper from the enzyme, as easier as lower the enzyme concentration.

5.21 Chemical Message

The ability of the antitumor drug to sequestrate Cu(II) suggests that, at least for the part of the metal ion involved in the coordination, the enzyme becomes less active in playing its role. This allows that a number of superoxide anions higher than in the absence of drug survive and enhance the formation of oxygen-centered radicals.

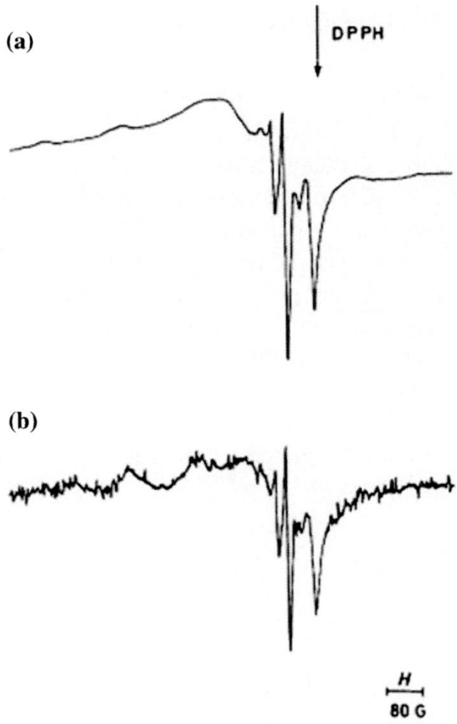

Fig. 5.16 ESR spectra in aqueous solution at pH 6.5 and −150 °C of **a** $CuCl_2$ doxorubicin and **b** Cu-SOD doxorubicin always r = 10 [14]

These radicals are killer of the human cells. It cannot be argued if the lowering of SOD activity in the presence of anthracycline is associated with all cells, both normal and tumor; however, it results in a destruction of tumor cells.

5.22 The Case of Generation of Free Radicals in Intact Tissues

The spectroscopic studies until now reported mainly concern molecular systems, these being models for the behavior of real systems [15]. The present case looks instead at the process of radicalization in the tissues of living organisms, as a consequence of the interaction with different drugs. In particular, considerable attention has been devoted to the free radical generation in myocardium and to its possible role as mediator of a variety of spontaneous and drug-induced pathologic conditions. Due to the difficulty in detecting free radicals, only indirect evidences of the toxic activity of these radicals have been collected. It became mandatory to find a method for the direct detection of radicals in tissues.

5.23 Discussion of the Case

Spontaneously beating atria were isolated from rats and suspended in a suitable bath solution. Heart rate and contractile force were continuously measured, discarding those showing arrhythmia. Suitable pieces of the intact organs were inserted into ESR tube for the analysis without previous manipulation. The spectrum of these samples is reported in Fig. 5.17a and compared with that of the same pieces of rat atria, pulverized under liquid nitrogen (Fig. 5.17b). The spectra obtained simply by inserting the pieces of the beating organs between the magnet expansions into the resonant cavity, without any manipulation, showed the presence of a very weak signal at g = 2.00. Pulverized atria showed instead a much more intense signal at the same resonant magnetic field.

These results suggest that the presence of paramagnetic centers is fully due to artifacts generated by the grinding of tissues.

(a)

(b)

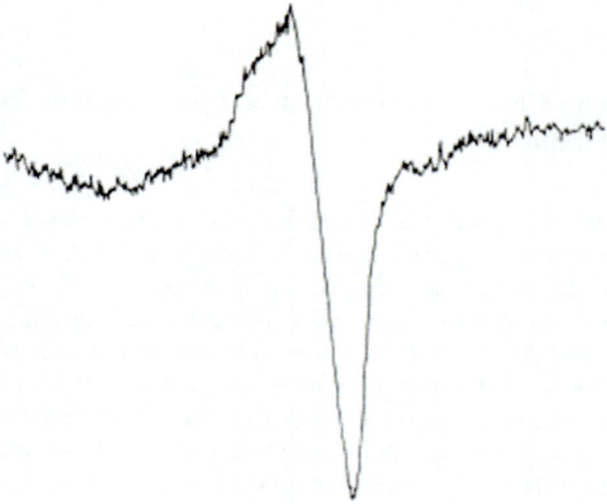

Fig. 5.17 **a** ESR spectrum of rat atria frozen in liquid nitrogen and **b** the same while pulverized under liquid nitrogen [15]

5.24 Chemical Message

In 1988, it became relevant to look for the detection of free radicals in tissues, thinking that the diseases associable to radicals could be thus controlled. The above-reported research strongly indicates that the tissue manipulation could artifactually generate radicals, simply due to mechanical break of the same tissues. It has been not excluded that the artifacts can also depend on the chemical treatment of tissues, like chemotherapy, but they are not generated by a chemical interaction effect.

5.25 The Case of Functional Defects in Semiconductor Oxides: ZnO

The semiconducting properties of several oxides, e.g., ZnO, have been under investigation for a long time [16]. A marked variation of the number of electrons in the conduction band has been brought about by modifying the surrounding atmosphere. This behavior is relevant to the use as chemical sensor. ZnO belongs to the non-stoichiometric oxide family, due to its excess of metal. Interstitial zinc atoms or anionic O^{2-} vacancies could act as donor centers as their electrons have sufficient thermal energy to enter the conduction band. Temperature and vacuum are both able to change the stoichiometry of ZnO, modifying the original defects. Chemisorption of oxidizing or reducing gases at the oxide surface also induces defect modification as the chemisorbed molecules interact with the electrons of the conduction band. Thus, an understanding of the nature of the defects and of their reactivity is a fundamental requirement to understand the electron transfer mechanism at the solid gas interface and to drive the application of the oxide in sensor devices.

5.26 Discussion of the Case

The main defects in ZnO are interstitial Zn_i centers and oxygen vacancies. Both are shallow defects and are able to ionize, transferring their electrons to the conduction band. For this reason, any change in the ionizable centers generates change in the conductivity of the material. Electron spin resonance spectra recorded under inert atmosphere show two groups of signals (Fig. 5.18): the first one with g values $g_A = 1.955$ and $g_B = 1.958$ and the second with g_C at 2.008 and 2.002. Signal A cannot be observed at room temperature, while it becomes strong at low temperature. Thermal treatment under vacuum leads to an increase of signal B, while A decreases. These results suggest that the signal A is attributable to zinc centers singly ionized, Zn^+, whose spin–orbit coupling contribution obliges to lower the

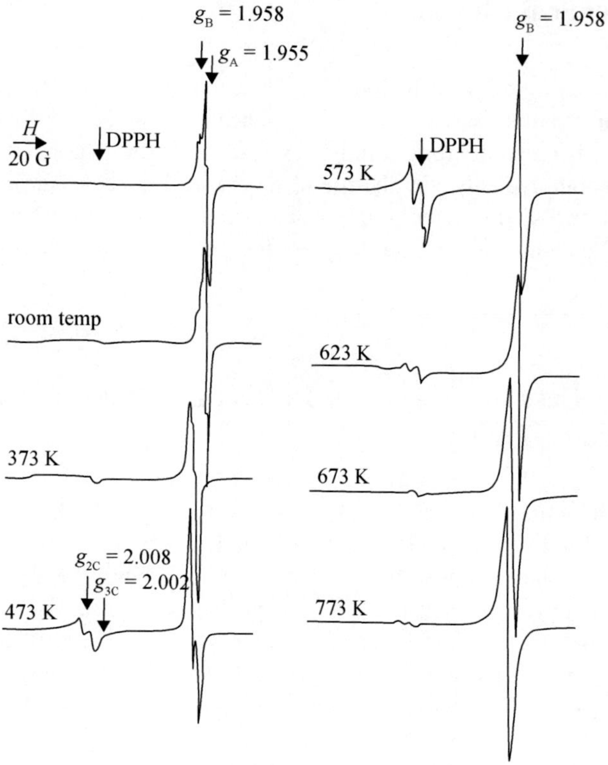

Fig. 5.18 ESR spectra of the pristine ZnO and of ZnO vacuum thermally treated at the indicated temperatures. The signals C are attributable to the perpendicular component of a rhombic species [16]

temperature to see the signal. The signal B is attributable to ionized oxygen vacancies V_0; in fact, it increases with the thermal treatment. In total, the following equilibria can be suggested

$$Zn \leftrightarrow Zn^+ + e^-$$
$$O^{2-} \leftrightarrow 1/2O_2 + V_0$$
$$V_0 \leftrightarrow V_0{}^+ + e^-$$

And they are interdepending.

It appears that varying the oxygen pressure the relative intensity of signals is varying. Under oxygen atmosphere, the resonance C becomes strong, due to formation of Zn^+-$O_2{}^-$.

5.27 Chemical Message

The oxygen vacancies dominate the total of defects in ZnO, and they are responsible for increase or decrease of conductivity as their amount increases or decreases.

For the first time, the defects in semiconductors, though ionized, became directly visible and could be associated with the physical properties of the oxide.

5.28 The Case of Functional Defects in Semiconductor Oxides: SnO_2

As well as ZnO, SnO_2 is one of the most common commercially materials used in gas sensor devices [17]. Although SnO_2-based gas sensors are widely used, their improvement in sensitivity and selectivity passes through a better knowledge of the mechanism by which they function.

The conductivity, whose variation is at the origin of the SnO_2-based gas sensors, has been postulated to be controlled by the Schottky barriers, through interaction with surface chemisorbed species. Neither interaction with lattice centers nor the consequent formation of donor centers has been postulated at the origin of the conductivity variations.

The scientific aim of this study is to assess whether SnO_2 conductance is controlled by Schottky barrier height and whether new donor centers become active.

5.29 Discussion of the Case

5.29.1 *Air Interaction with SnO_2*

When SnO_2 was pretreated in dry air at 673 K, no signals have been observed in the ESR spectrum. If the air is moist, strong resonance lines appear at g = 1.89 (Fig. 5.19). The resonances could be attributed to the creation of oxygen vacancies following the reaction

$$H_2O + 2Sn_{Sn} + O_0 \leftrightarrow 2(HO{-}Sn_{Sn}) + V_0$$

$V_0 \leftrightarrow V_0{}^+ + e^-$ where the paramagnetic species is $V_0{}^+$

Sn_{Sn} and O_0 represent Sn^{4+} and O^{2-} in regular sites and V_0 a neutral oxygen vacancy.

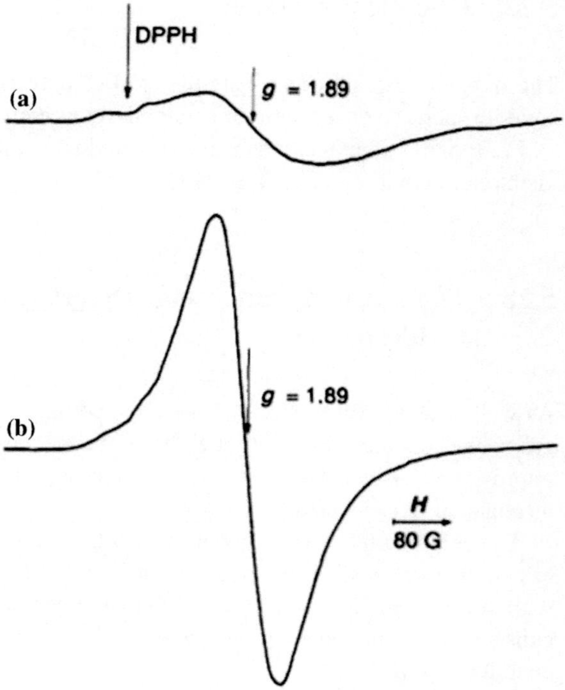

Fig. 5.19 ESR spectra recorded at 123 K of SnO_2 treated at 473 K under **a** wet air stream and **b** wet argon stream [17]

5.29.2 *Argon Interaction with SnO_2*

At variance as under air, in dry conditions, oxygen vacancies are formed, indicating that the equilibrium

$1/2O_2 + V_0 \leftrightarrow O_0$ is left-hand-shifted. Then, $V_0 \leftrightarrow V_0{}^{+} + e^{-}$.

5.29.3 *H_2 and CO Interaction with SnO_2, in the Presence of Air*

The interaction with dry H_2 produces a very large number of $V_0{}^{+}$. The effect, higher than that seen in moist atmosphere, is due to the following reactions

$$H_2 \leftrightarrow 2H^{\cdot}$$

$$Sn_{Sn} + O_0 + H^{\cdot} \leftrightarrow HO - Sn_{Sn} + V_0$$

$$V_0 \leftrightarrow V_0{}^{+} + e^{-}$$

Carbon monoxide interaction does not produce any effect in the absence of moisture. Under moist air, a number of vacancies were produced following the reactions

$$H_2O + Sn_{Sn} + O_0 + CO \leftrightarrow HCO\text{-}O_0 + HO\text{-}Sn_{sn}$$
$$Sn_{Sn} + HCO\text{-}O_0 \leftrightarrow Sn_{Sn} - HCO_2 + V_0$$
$$V_0 \leftrightarrow V_0^{\ +} + e^-$$

Figure 5.20 describes the ESR spectrum of the oxygen vacancies $V_0^{\ +}$ and of the superoxo derivative $Sn^{4+}\text{-}O_2^{\ -}$.

The resonance lines at g = 1.89 are attributable to the oxygen vacancies, under reducing CO/argon atmosphere (Fig. 5.20), while the spectrum of $Sn^{4+}\text{-}O_2^{\ -}$ $g_1 = 2.024$, $g_2 = 2.009$, $g_3 = 2.004$ becomes evident when O_2 was introduced. This assesses the possibility of understanding the structural variation at the SnO_2 surface that parallels the variations in conductivity. If toxic gases are mixed to air, being these mainly reducing gases, the conductivity variations are indicative of their

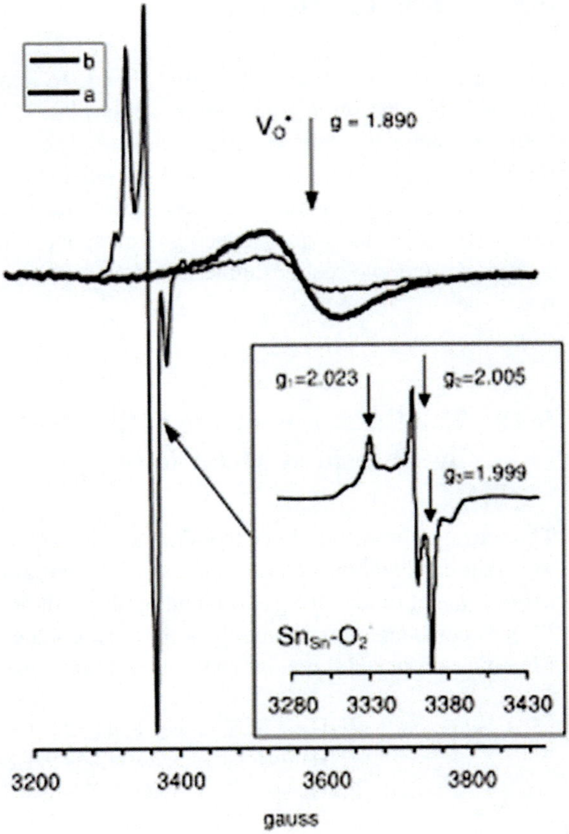

Fig. 5.20 ESR spectra of the paramagnetic species under **a** (stronger line) reducing and **b** oxidizing atmosphere [17]

presence in the atmosphere. See the case of nitrogen monoxide, NO interacting with SnO_2 surface. Specifically, several reactions can take place at the SnO_2 surface injecting electrons to the conduction band:

$NO \rightarrow NO^+ + e^-$ $\quad NO + O_0 \rightarrow NO_2^- + e^-$
$NO + O^- \rightarrow NO_2^-$ $\quad NO_2^- \rightarrow NO_2 + V_0^+$ (ionized oxygen vacancy)
$NO_2^- \rightarrow NO_2 + e^-$
$NO + O_2^- \rightarrow NO_3^-$
$NO_3^- \rightarrow NO_2 + O^-$

All increase the conductivity.
Other reactions consume the electrons decreasing the conductivity.

$O_2 + V_0^+ + Sn_{Sn} \rightarrow Sn_{Sn} - O_2^-$ (Sn_{Sn}) means the tin in a site of tin
$O_2 + e^- + Sn_{Sn} \rightarrow Sn_{Sn} - O_2^-$

5.30 Chemical Message

The variation of conductivity in SnO_2 is due to removal of oxygen from the oxide lattice. The operative temperature is about 500 °C, and the moisture is needing to operate in air as it creates oxygen vacancies. The results gained deeper insight when the SnO_2 samples were prepared by solgel method leading to highly purity samples.

It appeared convenient to perform gas contact in two distinct operative conditions: (i) under inert and reducing atmosphere, and (ii) under air stream. This reproduces the two realities under what the material for sensor devices can become operative.

5.31 The Case: Enhancing of the Oxide Sensing Properties by Transition Metal Ions

The sensing properties of semiconductor oxides can be enhanced by transition metal ions that catalyze the surface reactions, becoming a key point in the research concerning metal oxides for gas sensing applications [18]. Among these metal oxides, WO_3 is considered one of the most promising materials for ammonia gas detection. Literature proposed three competitive reactions for NH_3 oxidation on metal oxides

$$\begin{aligned} 2NH_3 + 3O_0 &\rightarrow N_2 + 3H_2O + 6e^- \\ 2NH_3 + 4O_0 &\rightarrow N_2O + 3H_2O + 8e^- \qquad \textbf{(R)} \\ 2NH_3 + 5O_0 &\rightarrow 2NO + 3H_2O + 10e^- \end{aligned}$$

where O_0 is an oxygen in a regular site of oxygen. In the presence of doping transition metal ions, the gas sensitivity increases; however, a direct inspection of the electronic configuration of the metal and of its relation with the activation mechanism is needed.

5.32 Discussion of the Case

Both Cr(III) centers and Cr(V) centers are paramagnetic and visible in the ESR spectrum. The following spectra show the behavior of Cr-doped WO_3 samples, containing different Cr amounts and thermally treated at different temperatures (Figs. 5.21 and 5.22).

The ESR signals of the as-prepared samples show at g = 1.98 the symmetrical lines of $Cr(H_2O)_6^{3+}$ with the metal in a high spin d^3 (signal β). This signal modifies in annealed samples, where dehydration changes $Cr(H_2O)_6^{3+}$ into Cr_2O_3-.

The removal of water from the Cr(III) coordination sphere was also investigated by optical analysis (Fig. 5.23). Although the diffuse reflectance spectra are dominated by the WO_3 gap transition around 450 nm, the as-prepared samples show two shoulders at 603 and 470 nm, ascribed to Cr(III) d-d transition in octahedral symmetry. The bands shift to 635 and 500 nm after annealing at 400 °C, and 645 and 514 nm after annealing at 700 °C. This agrees with the replacement of one or

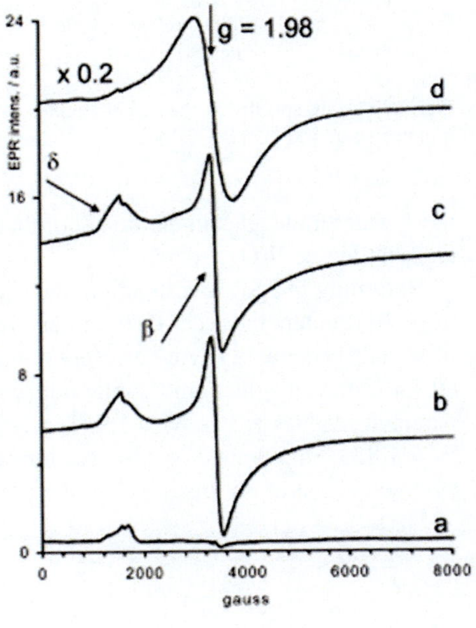

Fig. 5.21 ESR spectra recorded at 123 K of Cr-WO_3, as prepared, for increasing Cr contents. **a** 0.2%, **b** 1%, **c** 2%, **d** 5% [18]

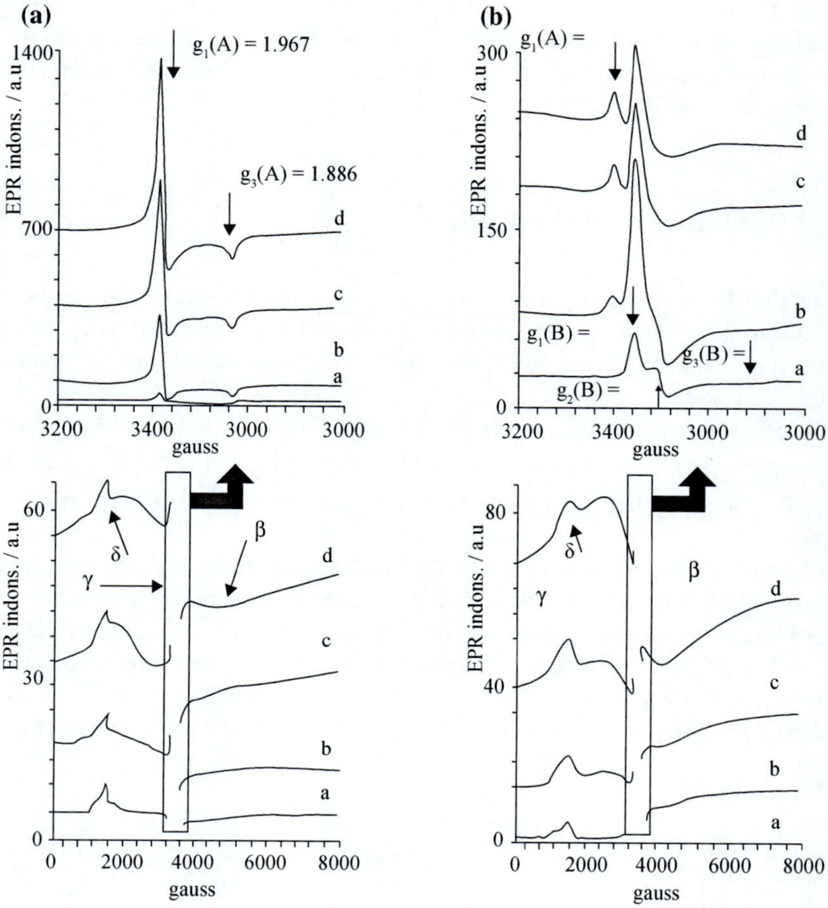

Fig. 5.22 ESR spectra of the same samples as in Fig. 5.21, while **a** annealed at 400 °C and **b** annealed at 700 °C [18]

more water molecules in the first coordination sphere of Cr(III) by a weaker ligand, e.g., the lattice WO_3 oxygen.

Returning to ESR investigation, the signal δ in Fig. 5.21, g = 3.5, is attributable to Cr(III) centers in a very distorted crystal field. In annealed samples, a new series of signals become evident, corresponding to a low-spin Cr species, whose g values all indicate a d^1 electronic configuration in axial or rhombic field. The magnetic parameter values are $g_{\|}(A) = 1.967$, $g_{\dagger}(A) = 1.886$, $g_1(B) = 1.949$, $g_2(B) = 1.912$, $g_3 = 1.823$, and both species are attributable to isolated Cr(V) centers whose symmetry is strongly distorted by the presence of $[Cr = O]^{3+}$ groups (chromyl).

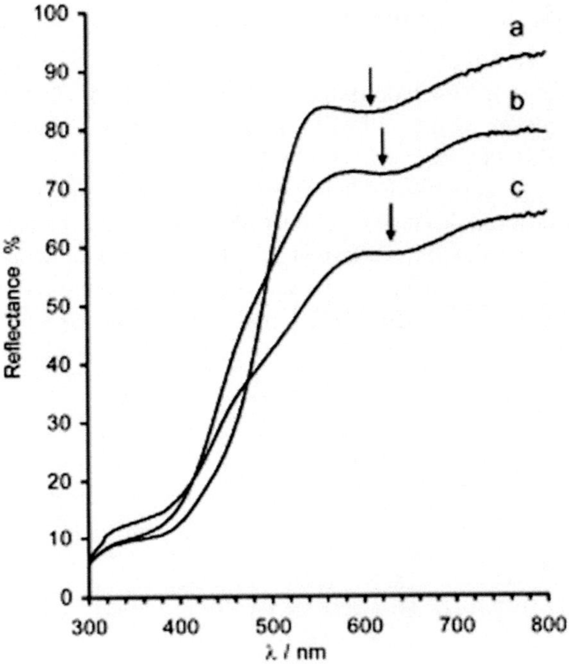

Fig. 5.23 Diffuse reflectance spectra of Cr-WO_3 samples **a** as prepared, **b** treated at 400 °C, **c** treated at 700 °C [18]

5.33 Chemical Message

It can be suggested that Cr(V) centers present in all the annealed samples are able to catalyze the oxidation of ammonia, by capturing the electrons deriving from reactions **(R)**; they are active in particular as they belong to a chromyl species emerging from the surface.

5.34 The Case of the Inhibitory Effect of Cu(II) Toward Protein Kinase C

Some trends in the development of new anticancer agents indicate a possible therapeutic goal, as an alternative to DNA damage, in the inhibition of specific cell functions, such as transduction pathways, triggered by growth factor receptor or by tumor promoter molecules [19]. On this purpose, the chelation of transition metal ions by anthracyclines, which appears to enhance the inhibitory action exerted by these drugs on protein kinase C (PKC), has been investigated, to assess the existence of an interaction between the enzyme and the drug and to suggest a molecular mechanism for the inhibitory effect.

5.35 Discussion of the Case

Figure 5.24 reports the inhibition effects of the copper centers, the copper complexes of the anthracycline epidoxorubicin, [(EpiDXR)] and [Cu(EpiDXR)$_2$], on PKC activity.

It appears that all three interacting agents inhibit PKC in a dose-dependent fashion, the strongest effect being shown by the [Cu(EpiDXR)$_2$] complex. Thus, it is mandatory to investigate the PKC interaction with Cu(II) and [Cu(EpiDXR)$_2$]. To this aim, being Cu(II) an ESR active center, the ESR spectra of Cu(II); [Cu (EpiDXR)$_2$]; Cu in PKC; and [Cu(EpiDXR)$_2$] + PKC were taken (Fig. 5.25).

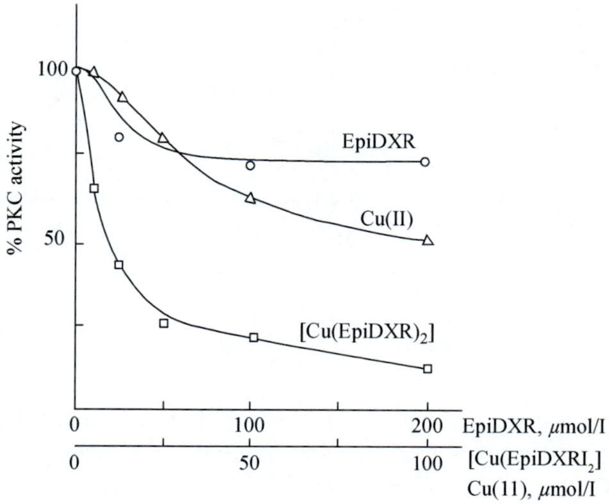

Fig. 5.24 PKC inhibition effects by epidoxorubicin (EpiDXR), Cu(II), and [Cu(EpiDXR)$_2$] [19]

Fig. 5.25 ESR spectra of **a** Cu(II), **b** [Cu(EpiDXR)$_2$], **c** Cu(II) in the presence of PKC, and **d** [Cu(EpiDXR)$_2$] in the presence of PKC [19]

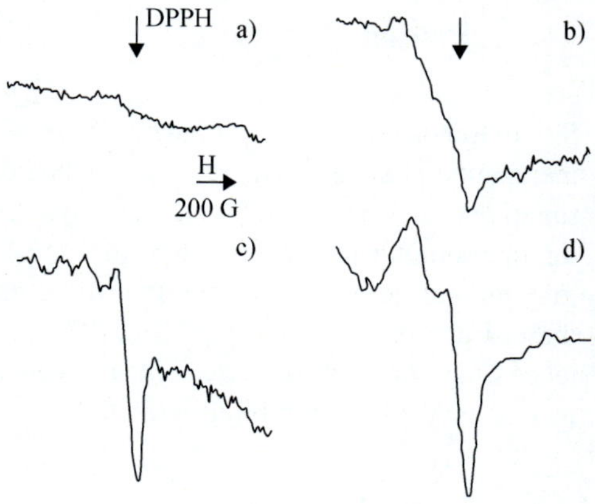

The spectra demonstrate that Cu(II) coordinates with the anthracycline and due to this coordination the metal ion becomes more suitable to be bonded by PKC. For this reason, the PKC inhibition in the presence of the anthracycline is higher.

5.36 Chemical Message

The geometric planar asset of [Cu(EpiDXR)$_2$] drives the PKC to bind copper, probably in a regular coordination site. This originates a stable bonding with the enzyme.

5.37 The Case of the Catalytic Oxidation of Propenoidic Phenols

Propenoidic phenols are constituents of lignin, the most abundant natural polymer after cellulose [20]. Wastewaters from the paper industry and agroindustrial activities contain a high concentration of polyphenols. These compounds are usually removed by oxidation with nitroaromatics, by alkaline oxidation or by electrochemical oxidation. It has proposed (Fig. 5.26) the oxidation with dioxygen in the presence of NN′ethylenebis (salicylideneiminato)cobalt (II) (CoSalen) as a catalyst. Two main degradation products, the 4-hydroxybenzoic acids and the commercially important 4-hydroxybenzaldehydes, were obtained. For this reason, the activation of dioxygen by CoSalen in the presence of propenoidic phenols received deeper interest. Compound 1 showed the highest activity in chloroform solution, good activity in methanol, and very low activity in pyridine. Thus, the oxidative catalysis is favored in acidic solution. Which is the reason?

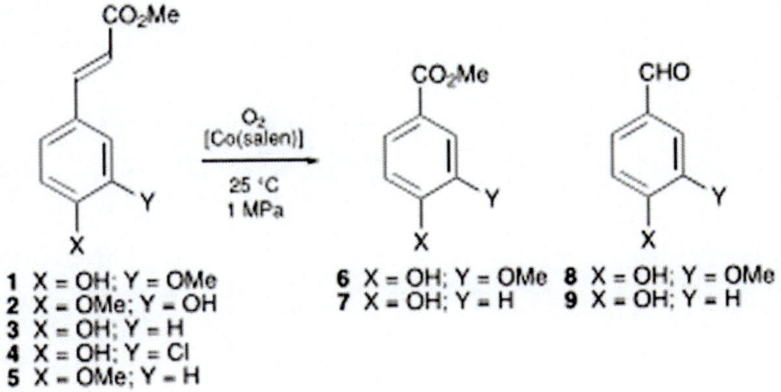

Fig. 5.26 Hydroxybenzoic acids reacted with O$_2$(1–5)

5.38 Discussion of the Case

The investigation, essentially made by ESR, begun from the spectrum of CoSalen in the presence of E-methyl ferulate (**1**) (Fig. 5.27) in deaerated chloroform solution.

The contact with oxygen in the presence of (**1**) modifies the resonances of CoSalen and gives rise to the formation of the superoxo derivative of CoSalen and of a radical species (Fig. 5.28).

It has been suggested that in the presence of O_2 the following processes take place

$$[Co^{II}(salen)] + ROH + O_2 \rightleftharpoons [Co^{III}(salen)(ROH)(O_2)]^-$$
$$[Co^{III}(salen)(ROH)(O_2)]^- + ROH \rightleftharpoons [Co^{III}(salen)(ROH)(RO^*)] + HO_2^-$$
$$[Co^{III}(salen)(ROH)(RO^*)] + HO_2^- \rightleftharpoons [Co^{III}(salen)(RO^-)(RO^*)] + H_2O_2$$

The formation of the phenoxy radical Co^{III}(Salen)(ROH)(RO$^\cdot$) is the determining step of the whole oxidative process, and it is mediated by the superoxide anion. The free coordination of (**1**) to the cobalt is needed to the process.

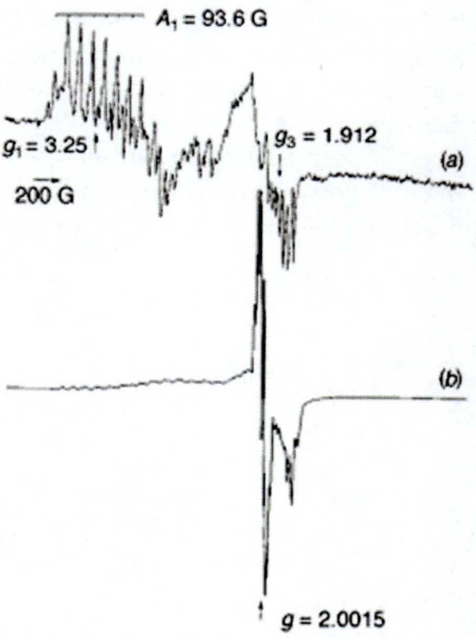

Fig. 5.27 **a** ESR spectrum of CoSalen in $CHCl_3$ in the presence of (**1**) under inert atmosphere and **b** under oxygen atmosphere [20]

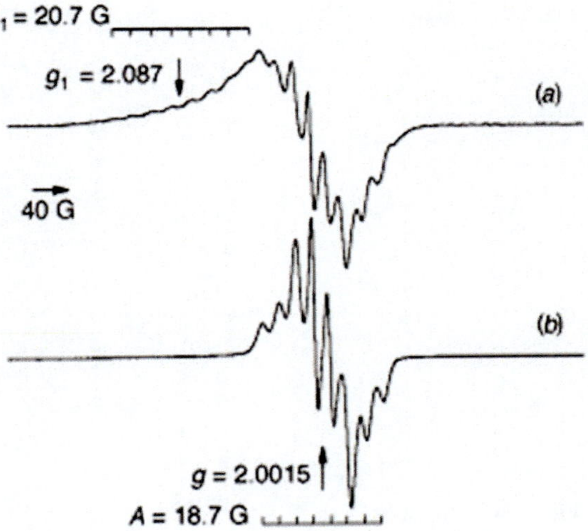

Fig. 5.28 ESR spectra of CoSalen in $CHCl_3$ in the presence of (**1**) contacted with O_2, **a** 5 min and **b** 15 min [20]

5.39 The Case of Sn-Doped Solgel Silica Glasses

Silica materials have been extensively studied as for the possibility of designing the change of their light-induced refractive index [21]. This aims to prepare glasses with different optical properties, to be used in optical fibers. Photorefractive properties are generally associated with the defects related to silica-doped, typically germanium-doped, that means three-coordinated Ge centers with one unpaired electron in an sp^3 orbital. It is intended that the dopant element must be homogeneously dispersed in silica and located in substitutional position. Looking for dopants different from Ge, a good candidate seemed to be tin, Sn basing on the large increase in refractive index it showed co-doped with Ge. The solgel method of synthesis has been used, as it allows to obtain Sn-doped silica glasses by a low-temperature method, from the corresponding element molecular precursors (alkoxides); instead, high temperature or melting procedure of the related oxides leads to segregation of the dopant element. The optical absorption properties of glasses have been studied by UV–visible absorption spectroscopy; the location of Sn atoms has been checked by electron spin resonance.

Figure 5.29 shows the silica glasses with different Sn contents.

5.40 Discussion of the Case

Figure 5.30 shows the ESR spectra of Sn-doped silica glasses, after X-ray irradiation able to generate dangling bonds in the glass structure. ESR spectra show the signals of the E′ Sn centers that are three-coordinated Sn centers with one unpaired electron in an sp^3 orbital. These have orthorhombic symmetry field with

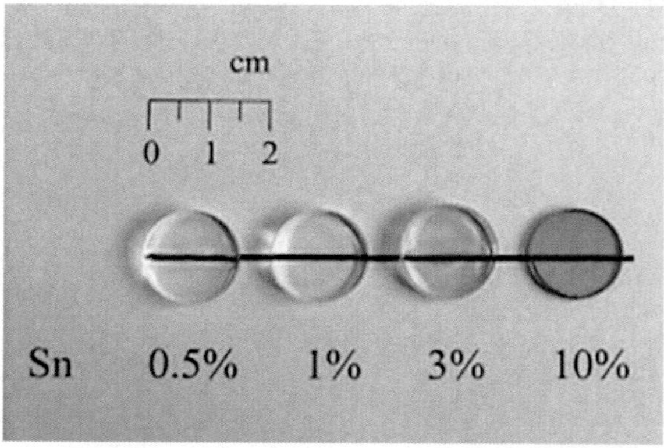

Fig. 5.29 Sn-doped silica glasses [21]

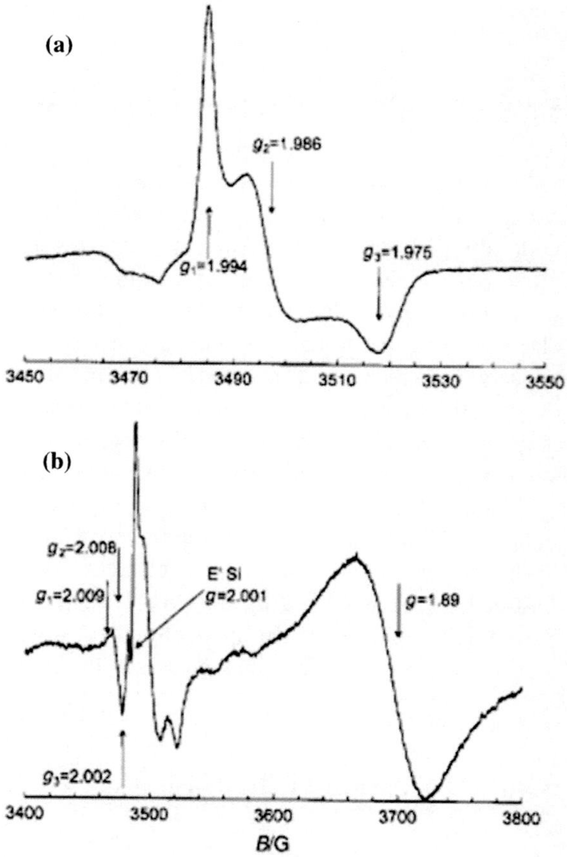

Fig. 5.30 ESR spectra of Sn-doped silica glasses, after X-ray irradiation. **a** Sn 0.5% and **b** Sn 10% [21]

$g_1 = 1.994$, $g_2 = 1.986$, and $g_3 = 1.975$; moreover, it was seen the narrow line of E′ Si centers at $g = 2.001$ and the resonances of non-bridging oxygen hole centers ≡ Si-O˙ and peroxy radicals ≡ Si-O-O˙ at $g_1 = 2.009$, $g_2 = 2.008$, and $g_3 = 2.002$. In the glass with high Sn content, color like dark yellow, it appears also the resonance of the singly ionized oxygen vacancies of crystalline SnO_2. The electron spin resonance spectra revealed thus a fantastic method to distinguish well-diluted Sn-doped silica glasses, from those which contain segregated nanoparticles of crystalline SnO_2. These nanoparticles can be recognized by an absorption at 340 nm, corresponding to the energy gap of SnO_2, and also show the ionized oxygen vacancies of the SnO_2 lattice at $g = 1.89$.

5.41 Chemical Message

By using the solgel procedure, essentially starting from solutions of the element precursors, elements different from the oxygen, silicon can be substituted by Sn and this leads to obtain homogeneous glasses with different amounts of dopant. Then, it results in the possibility of designing glasses with various and controlled refractive indices to be used in projecting optical fibers. The success of the synthesis is controlled by the physicochemical characterization of the defects induced by the dopant element.

5.42 The Cases of Photocatalytic TiO_2

Titanium dioxide is widely used in different applications, where the most relevant is the photocatalysis (induced by UV irradiation) [22]. TiO_2 exists in different crystalline forms: anatase, rutile, and brookite. These are expected and have been verified to have a different catalytic activity; nevertheless, the reasons of the differences are still object of investigation. In fact, while some authors claimed to the higher surface area or porosity, as origin of higher efficacy, others suggested that the difference in surface defects definitely decides the activity.

It only remains to select specific crystalline TiO_2 phases and investigate the location of the defects. Defects in TiO_2, generated by UV irradiation, consist of holes (electron vacancies in the valence band of TiO_2) (h^+) and free electrons (e^-) in the oxide conduction band. They have in charge the photomineralization process, following the reactions here reported.

$$TiO_2 + h\nu \rightarrow e^- + h^+$$
$$e^-_{CB} + H_2O_2 \rightarrow OH^{\cdot} + OH^-$$
$$e^-_{CB} + O_2 \rightarrow O_2^{\cdot -}/HO_2^{\cdot}$$
$$h^+ + OH^- \rightarrow OH^{\cdot}$$
$$OH^{\cdot} + S\,(\text{organic substrate}) \rightarrow CO_2 + H_2O$$

Let us consider that e^-_{CB} are actually located on Ti^{3+} centers or O_2^- centers and h^+ on O^- centers. All these species are paramagnetic.

5.43 Discussion of the Case

The investigation concerning the above-mentioned electronic defects has been done by studying the mineralization of phenol in the presence of both pure (anatase and rutile) and mixed phase TiO_2 (see phase compositions in Table 5.4), employing as oxidative agents both H_2O_2 and O_2.

It has been found that in the case of H_2O_2, rutile particles having larger dimensions and higher aspect ratio displayed the highest catalytic activity, probably due to the lower tendency of electrons and holes to recombine. By using O_2, rutile displays lower efficacy probably as the oxygen preferentially chemisorbs at the surface of the nanosized anatase particles and acts as electron scavenger, inhibiting the electron–hole recombination. These hypotheses have been investigated in situ by electron spin resonance.

Electron and hole trap centers generated by UV irradiation could be related to the charge carrier life time and to the catalyst efficacy. Figure 5.31 shows that in mixed rutile–anatase samples the resonance of e^-_{CB} is very strong and persistent. This means that the associated defects are well separated in charge. Recombination is

Table 5.4 Phase composition of TiO_2 samples

Phase composition, BET-specific surface area, and XRD particle size of hydrothermal TiO_2 samples

	Phase composition		BET surface area ($m^2\ g^{-1}$)	XRD particle size	
	Anatase (wt%)	Rutile (wt%)		Anatase *d*(1 0 1) (nm)	Rutile *d*(1 1 0) (nm)
R100	0	100	22	–	53
R61	39	61	46	14	68
R57	43	57	60	15	31
R48	52	48	64	14	26
R20	80	20	68	16	35
R0	100	0	91	15	–

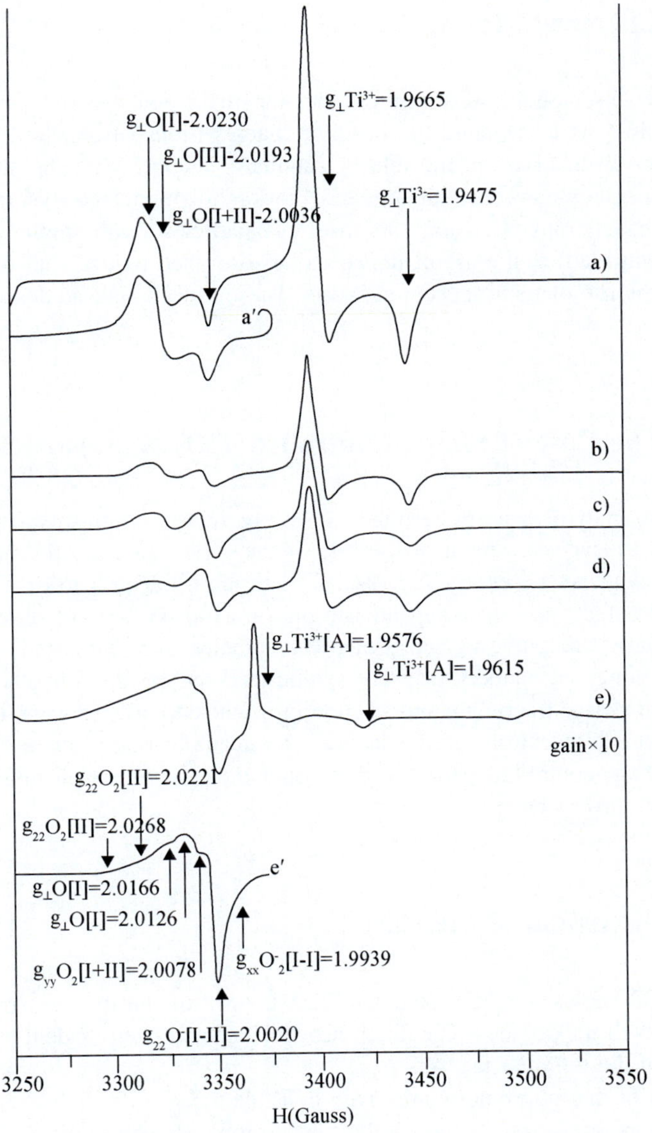

Fig. 5.31 ESR spectra recorded at 10 K of samples a R100, b R61, c R48, d R20, e R0; a′ and e′ are spectra simulations [22]

thus a slow process. In pure anatase sample, the defects are in a very small number suggesting that on anatase alone the recombination of $e^-{}_{CB}$ and Ti^{3+} is a favored process. When anatase and rutile are mixed, rutile accepts the electron from anatase and gives holes to the same. This is equivalent to give stability to the charge separation. In addition, anatase stabilizes O_2^- (line e′ in Fig. 5.31).

5.44 Chemical Message

The above-discussed results suggest that the main goal for developing of new photocatalysts is to separate the positive charges (holes) from the negative ones (electrons). Both of them are able to singularly interact with the substrate. The charge separation has constituted the basis for optimizing the catalyst materials, and the strategies ranged from the use of particles with high aspect ratio (shape-controlled) to the use of doped oxides or mixed oxides with different band gaps, to the use of non-innocent supports. These aspects will be described in next chapters.

5.45 The Case of Shape-Controlled TiO_2 Nanoparticles

The photocatalytic activity resulted is strictly related to the exposed surfaces; however, a clear assessment of the role of the crystal faces in the photocatalytic processes appears necessary [23, 24]. To this aim, the properties of the photogenerated defects and their dependence on the exposed faces of different anatase crystals have been investigated. Spherical particles, nanobars, and rhombic and rhombic elongated nanocrystal were synthesized by specific solvothermal procedures. The charge trapping centers active in irradiated particles were Ti^{3+} O^{-} O_2^{-}, all detectable by electron spin resonance. The quantification of these defects in the various shape-controlled particles gives an indication of which defects are prevailing on a given face.

5.46 Discussion of the Case

Figure 5.32 shows the TEM and HRTEM images of the different materials and indicates their nick names. The differences in shape are well evident as well as the location of the different facets.

Details of the morphology are given in Table 5.5.

It appears from the $t_{1/2}$ values that the rhombic-shaped particles are the most catalytically active. Thus, the photocatalytic activity is greatly dependent on the particle shape. Interestingly, the best performance occurs for R-shaped particles which also display a relatively high SSA_{BET}. Despite having the lowest SSA_{BET}, RE nanocrystals show relevant activity too. Finally despite the very high SSA_{BET}, NBs show the worst photoefficacy. These results suggest that the difference in surface area is not representative of the photocatalytic properties. Instead, the exposed crystal faces and their relative SSA_{BET} play the major role. Specifically,

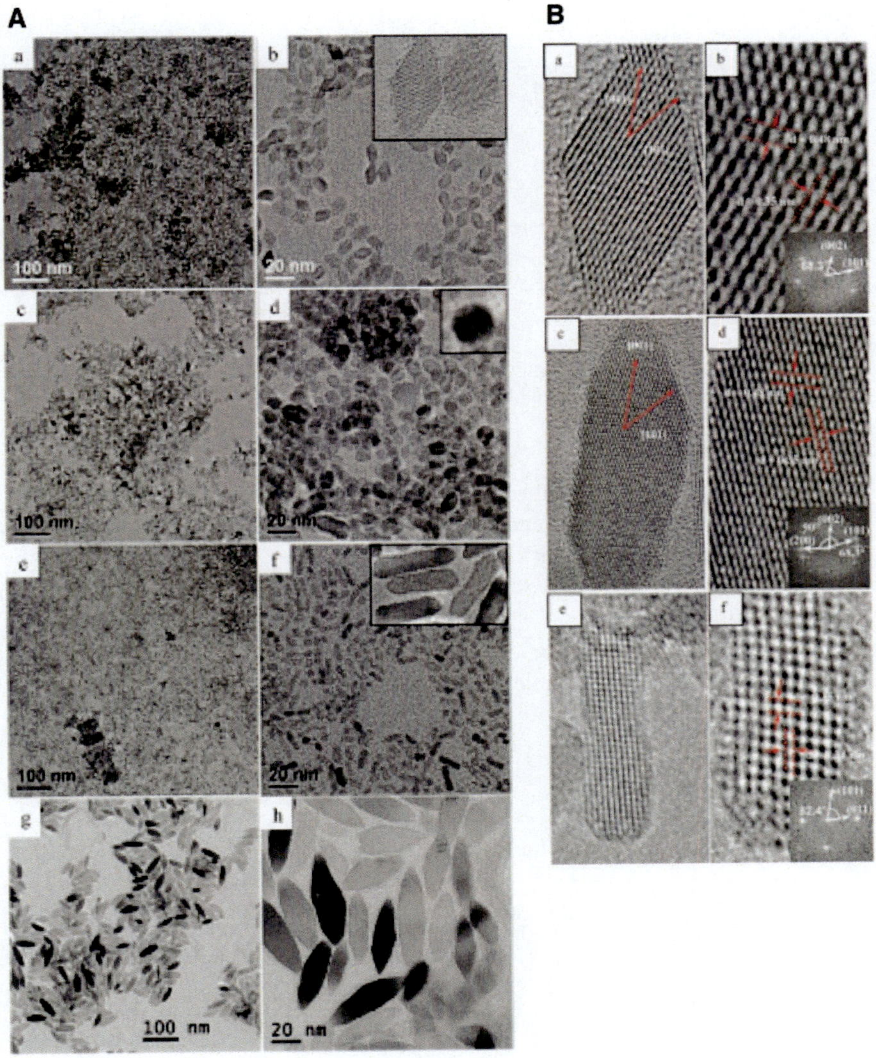

Fig. 5.32 **A** TEM images of the different shape particles: **a** and **b** rhombic (**R**); **c** and **d** spheric (**SP**); **e** and **f** nanobars (**NB**); **g** and **h** rhombic elongated (**RE**); **B** HRTEM images of the same particles **a** and **b** rhombic (**R**); **c** and **d** rhombic elongated (**RE**); **e** and **f** nanobars (**NB**) [23, 24]

the large SSA_{BET} of the exposed {001} faces and the relatively high area of the exposed {101} appear crucial in determining the remarkable activity of R and RE nanocrystals. Conversely, the low SSA_{BET} of exposed {001} faces and the high of {101} faces agree with the worst activity of NBs. SP nanoparticles have intermediate behavior.

Table 5.5 Morphologic characters of shape-controlled particles including the area of the exposed facets

Sample	L_{XRD} (nm)	Pore volume (DCPV, $cm^3\ g^{-1}$)	SSA_{BET} ($m^2\ g^{-1}$)	Exposed {001} crystal facets (%)	Exposed {101} crystal facets (%)	SSA_{BET} of exposed {001} crystal facets ($m^2\ g^{-1}$)	SSA_{BET} of exposed {101} crystal facets ($m^2\ g^{-1}$)	$t_{1/2}$ (min)
SP	7.6	0.47	178.8					130.0
NB	7.3	0.27	227.0	5.8	94.2	13.1	213.8	183.7
R	13.1	0.35	199.0	10.6	89.4	21.2	177.9	89.3
RE	16.5	0.21	170.5	9.2	90.8	15.7	154.8[a]	124.1

The photocatalytic performances in terms of half decay time are also indicated for the phenol oxidation
[a]Included the {010} and {100} minority facets

It can be hypothesized that the difference in the catalytic activity is depending on the rate of recombination between holes (h^+) and electrons (e^-) within the particles that is on the stable separation of the opposite charges. For this reason, the best pathway seemed to directly investigate these charges, which are both paramagnetic and ESR active.

We firstly report in Fig. 5.33 the key of identification of the paramagnetic defects, referred to the R particle shape.

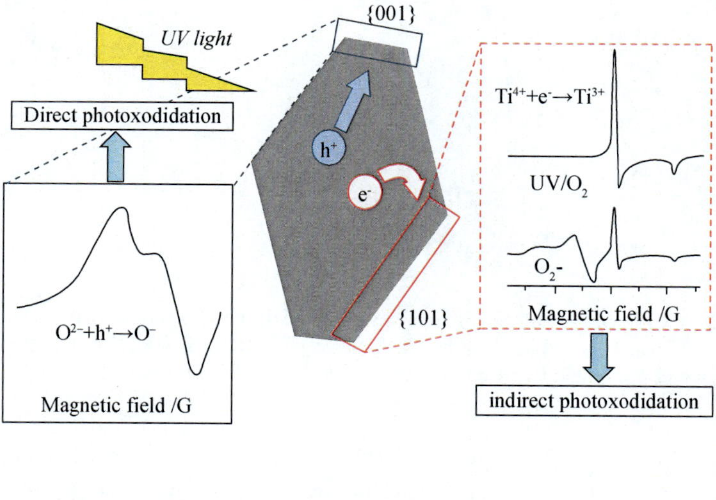

Fig. 5.33 Paramagnetic defects in a rhombic particle [23, 24]

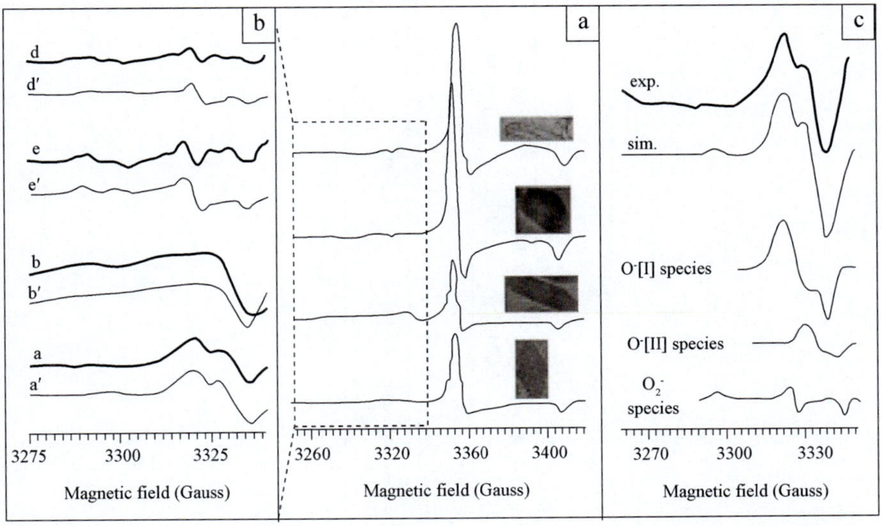

Fig. 5.34 ESR spectra of UV-irradiated different shaped nanoparticles. The shapes are inside part **a**. **b** Magnification of O^- signals **a–d**, with their simulation **a′–d′**. **c** Deconvolution of ESR signals into O^- and O_2^- for R nanoparticles [23, 24]

Well evident are the anisotropic resonances of O^- (holes associated with O^{2-} centers), those of Ti^{3+} (electrons associated with Ti^{4+} centers) and of O_2^- (electrons associated with O_2). In detail, Fig. 5.34 reports the ESR spectra for the different types of nanoparticles (Table 5.6).

In association with the previous data, it is reported (Fig. 5.35) the decay curves of the total organic carbon (TOC) detected for the phenol oxidation.

Summarizing all the results, it appears that the concentration of trapped hole (O^-) centers increases with increasing the {001} surface area and the photoactivity, while the amount of Ti^{3+} centers increases with the specific area of {101} facets and the highest value occurs for the sample with the worst photooxidative efficacy. {001} surfaces can be considered essentially oxidation sites, while the {101} provide the reductive sites. In the presence of oxygen, the Ti^{4+}-O_2^- species mainly located on {101} surfaces could indirectly contribute to the oxidative process.

Table 5.6 Magnetic tensor values of the species in the differently shaped particles, with the amount of the O^- species

Sample	O^- centers (%)	O^- species	Ti^{3+} species	O_2^- species
NB	2	$g_\perp = 2.0126$, $g_\parallel = 2.0046$	$g_\perp = 1.9867$ $g_\parallel = 1.9570$	O_2^-[I] : $g_{zz} = 2.0252$, $g_{yy} = 2.0080$, $g_{xx} = 1.9990$ O_2^-[II] : $g_{zz} = 2.0212$, $g_{yy} = 2.0080$, $g_{xx} = 1.9990$
SP	1	$g_\perp = 2.0129$, $g_\parallel = 2.0047$	$g_\perp = 1.9876$ $g_\parallel = 1.9579$	O_2^-[I] : $g_{zz} = 2.0254$, $g_{yy} = 2.0083$, $g_{xx} = 1.9990$ O_2^-[II] : $g_{zz} = 2.0215$, $g_{yy} = 2.0083$, $g_{xx} = 1.9990$
R	78	O^-[I] : $g_\perp = 2.0090$, $g_\parallel = 2.0001$ O^-[II] : $g_\perp = 2.0050$, $g_\parallel = 1.9975$	$g_\perp = 1.9866$ $g_\parallel = 1.9575$	$g_{zz} = 2.0258$, $g_{yy} = 2.0072$, $g_{xx} = 1.9965$
RE	44	$g_\perp = 2.0090$, $g_\parallel = 2.0004$	$g_\perp = 1.9866$ $g_\parallel = 1.9582$	$g_{zz} = 2.0280$, $g_{yy} = 2.0070$, $g_{xx} = 1.9994$

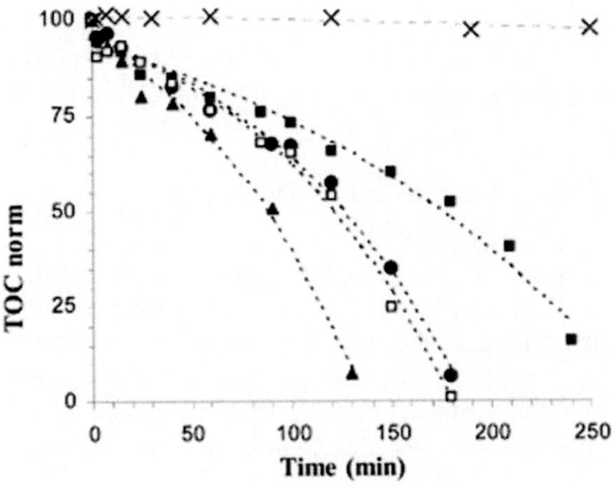

Fig. 5.35 Irradiation in the presence of O_2 of (×) blank, without catalyst, (■) NB, (•) SP, (□) RE, and (▲) R TiO_2 nanocrystals

5.47 Chemical Message

The conclusion of the literature that preceded this topic was that the photocatalytic activity depends on the rate of charge recombination within the semiconductor oxide. Instead, our results demonstrate that there is strong attitude of the different facets to give rise to defects, depending on the atom location. The recombination rate is more properly depending on the shape of the facet. Thus, the crystal symmetry, once again, decides the activity of the catalysts. The preparation of innovative photocatalysts could realistically go through the orientation of the active facets toward the fluid–solid interphase.

5.48 The Case of Charge Separation in TiO_2 Embedded in Membrane

The use of TiO_2 requires, either in liquid or in gas phase reactions, to provide an easy recovery of the catalyst [25]. Due to this reason, the powder oxide is not the most suitable material for industrial applications. Different matrices have been suggested for it, inorganic or polymeric, being careful that the host was able to fully retain the catalyst allowing complete passage of products and light.

Among the innovative matrices, an important role is played by polymeric membranes constituted by acrylates and metacrylates, copolymerized with other monomers, and embedding TiO_2 by a grafting process. These membranes are highly porous and permeable, allowing good interaction with the UV radiation and the reactants, also having good mechanical properties.

The present example shows not only that the inclusion of TiO_2 into a polymeric membrane (TPM) allows easy manipulation of the catalyst, but also that the catalyst is assisted by the membrane and by O_2 in keeping holes separated from the electrons. In particular, the interaction of holes with the polymeric membrane, in O_2, results in the formation of peroxy radicals, which can be considered responsible for the stabilization of both holes and electrons (trapped on Ti^{3+} centers).

5.49 Discussion of the Case

In Fig. 5.36 is reported the catalyst morphology.

Figure 5.37 reports the mineralization curves of phenol under UV irradiation, in the presence of O_2 and TPM catalyst.

It appears that the efficacy of the embedded catalyst is comparable with that of the slurry catalyst, as if the matrix would not affect the interaction between substrate and catalyst. For this reason, a more accurate investigation of the electronic

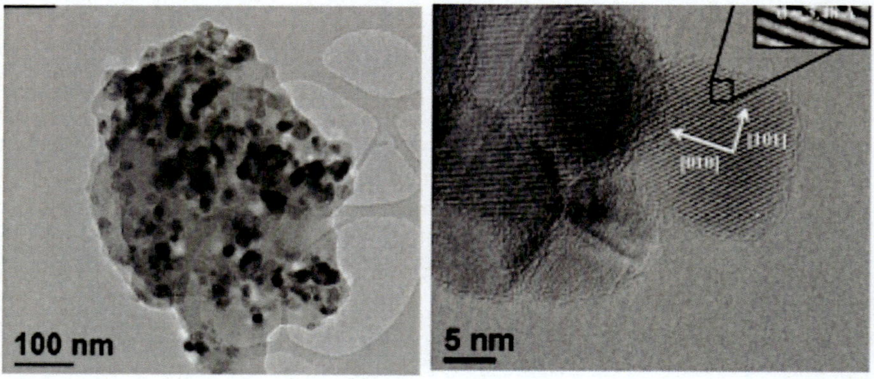

(a) (b)

Fig. 5.36 Anatase crystals embedded in the membrane. **a** TEM and **b** HRTEM images [25]

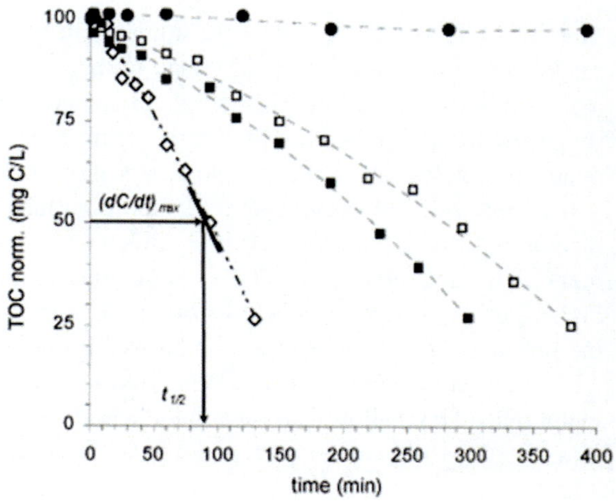

Fig. 5.37 Phenol mineralization curves obtained by the total organic carbon (TOC) analysis. ● In the absence of catalyst, □ after the first run of reaction, □ (fill the squares) after 6 runs, ◇ P25 slurry catalyst [25]

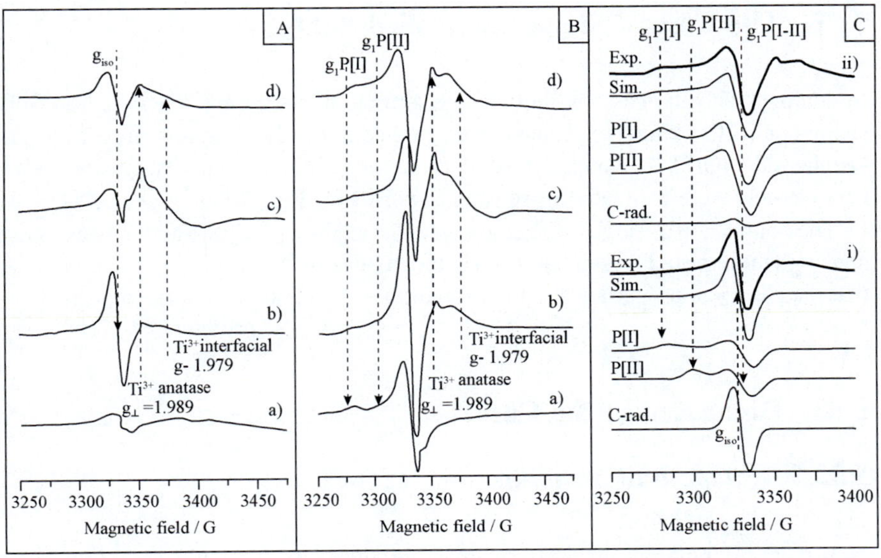

Fig. 5.38 ESR spectra under vacuum (**A**) and in the presence of O_2 (**B**) after 30 min of the photocatalytic run (line **b**) and after completion of the first (line **c**) and sixth run (**d**). The membrane without TiO_2 is also reported (line **a**), **C** deconvolution of the **B** signals after completion of the first (**i**) and sixth (**ii**) runs into peroxy species. Dashed lines guide the detection of the species [25]

properties of embedded TiO_2 became necessary. The well-known defects, holes, and electrons (Ti^{3+}) are ESR active and have been investigated.

After irradiation in vacuo, a radical species associated with the polymer matrix was present (Fig. 5.38 line a). During the runs (lines b, c, d), only the anatase resonance lines are visible. No rutile Ti^{3+} centers were ever detected, except the rutile–anatase interfacial ones.

Irradiation under oxygen leads instead to increase the radical amount in matrix; moreover, Ti^{3+} amounts located in anatase become higher (lines b–d). This suggests that the photogenerated holes are trapped by the polymeric matrix while the electrons are involved in the reduction of Ti^{4+} to Ti^{3+}.

The membrane plays the role of separating the charges, allowing that the electrons are trapped by Ti^{3+} centers and the holes give peroxy radicals (Fig. 5.38C). These later are responsible for the oxidation reaction.

5.50 Chemical Message

The inclusion of TiO_2 photocatalyst into a porous membrane, made by suitable polymers, allows that some of the bonds in polymers can be broken by the photogenerated holes and carbon-centered radicals trap oxygen molecules. This automatically separates the charges and generates highly oxidant species able to enhance the catalytic effect of the pristine TiO_2. Thus, modulating the membrane composition is possible to modulate the photocatalytic activity.

5.51 The Case of Shape-Controlled SnO_2

In analogy with the case of TiO_2, while aiming at a different functionality, SnO_2 shape-controlled nanocrystals have been studied as for their sensing properties [26]. Octahedral (OCT), elongated dodecahedral (DOD), and nanobar (NBA) nanocrystals were evaluated in sensing toward CO (Fig. 5.39).

The abundance of singly ionized oxygen vacancies $V_0{}^+$, ESR active, was associated with the exposed crystal faces and in turn to the sensing response. Results indicate that the electrical properties and the formation of V_0 centers are interconnected.

5.52 Discussion of the Case

5.52.1 *Shape and Structure*

Figure 5.40 reports the shapes of nanobar and dodecahedral particles to describe as the data can be read.

For OCT nanoparticles, Fig. 5.41 describes structure and orientations.

5.52.2 *Identification of Defects*

In order to study how the exposition of specific crystal faces affects the reactivity of the nanocrystals with different shapes, as concerns the reactivity of the species $V_0{}^+$ O^- O^{2-} $O_2{}^-$ involved in the sensing mechanism, a comprehensive ESR investigation of OCT, DOD, and NBA was carried out (Fig. 5.42).

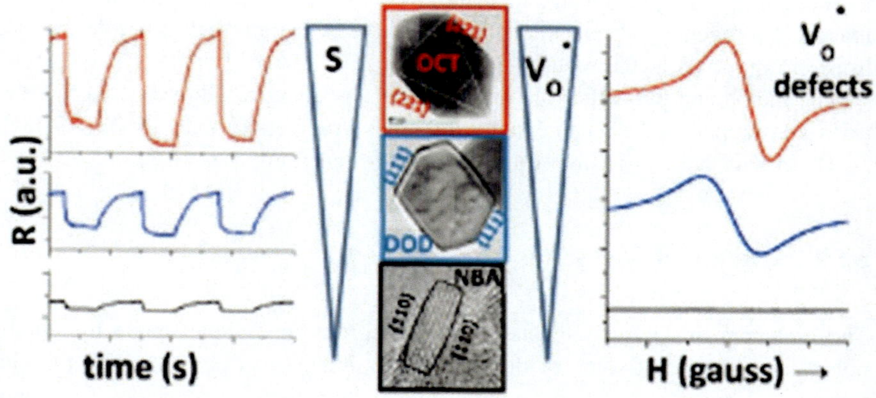

Fig. 5.39 Shapes, defects, and electric responses in SnO_2 nanoparticles [26]

Fig. 5.40 Geometrical model of **a** DOD nanocrystals exposing eight {111} and four {110} facets and **b** of NBA nanocrystals with {110} and four {001} facets [26]

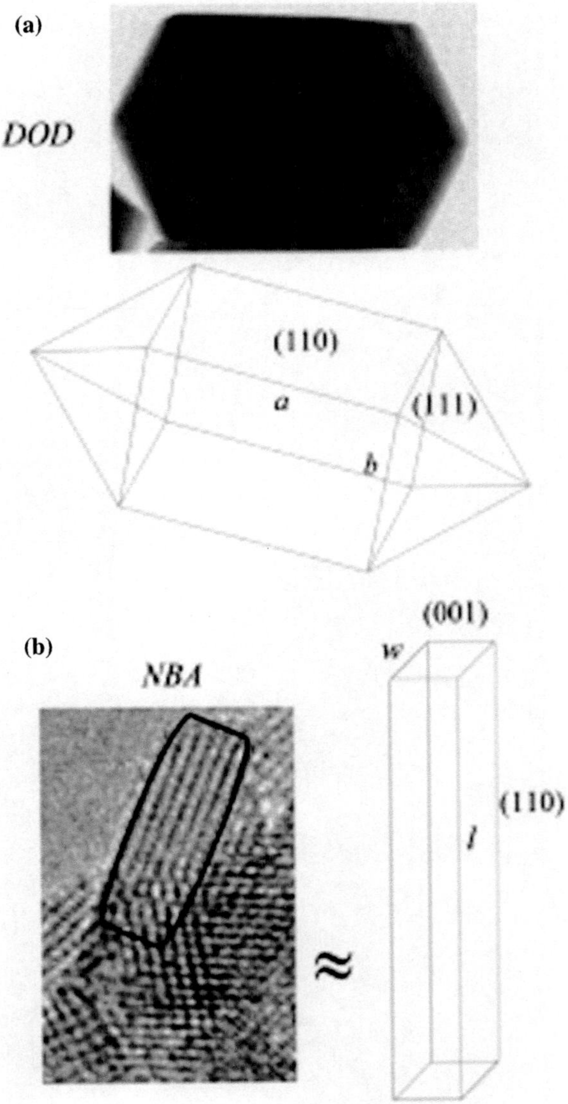

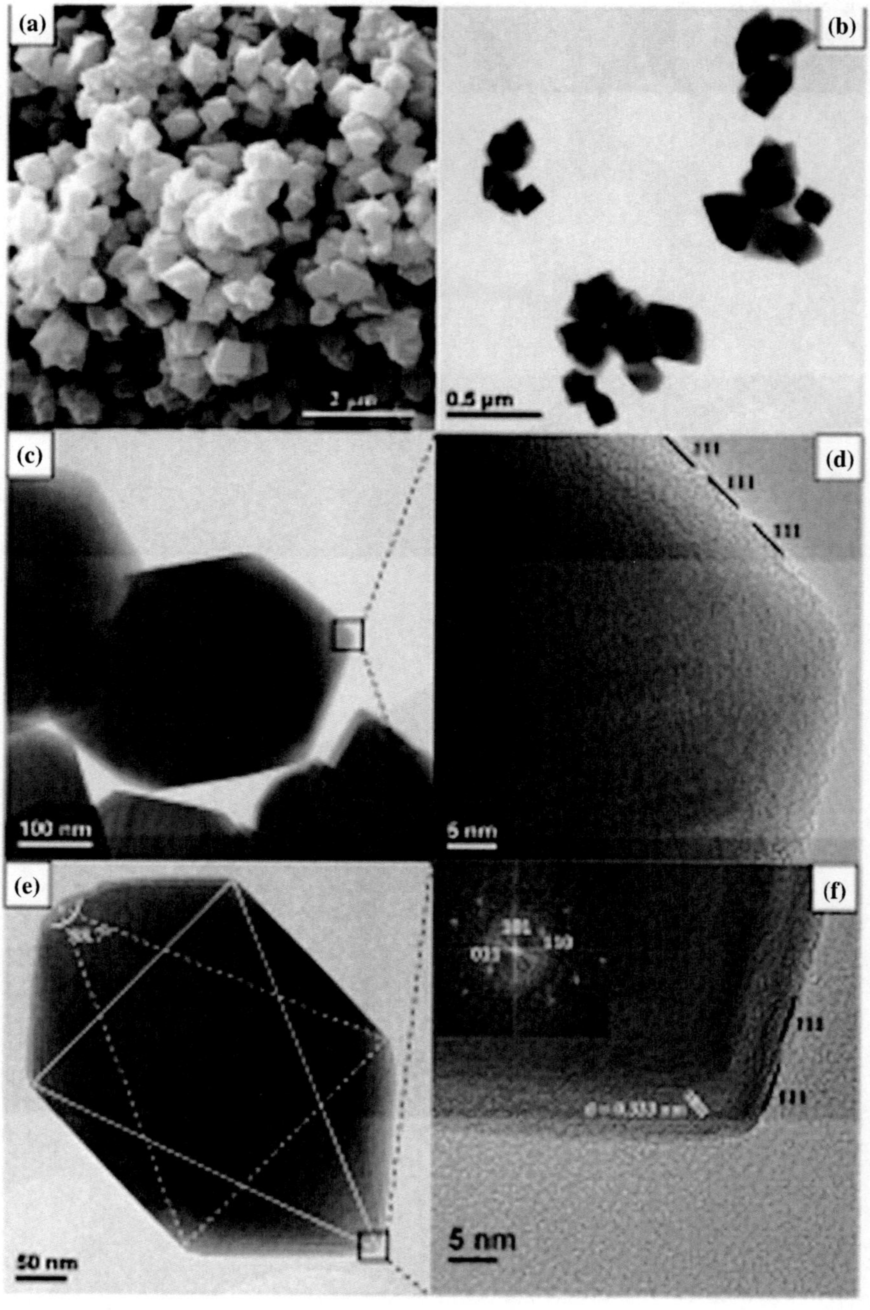

Fig. 5.41 Representation of OCT nanoparticles: **a** SEM, **b–d** TEM, **e**, **f** HRTEM; the inset in **f** is the crystalline structure detail of the octahedral nanoparticle [26]

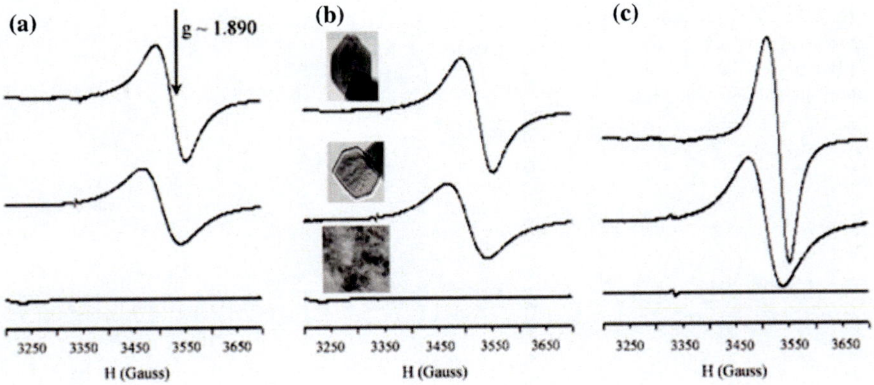

Fig. 5.42 **a** As-prepared sample, **b** after treatment under air stream, **c** after treatment with CO/argon. Insets in **b** show the corresponding shapes [26]

The spectra of the as-prepared OCT and DOD samples show the presence of resonances at g = 1.98, attributable to electrons trapped in singly ionized oxygen vacancies $V_0{}^+$. No signal was detected in NBA nanoparticles. Air does not affect the spectra, while under CO/Ar atmosphere the $V_0{}^+$ amount increases only for OCT and DOD nanocrystals. This suggests that new oxygen vacancies are created according to the following reactions.

$$O_0 + CO \rightarrow CO_2 + V_0$$
$$V_0 \rightarrow V_0{}^+ + e^-$$

The amount of paramagnetic oxygen vacancies is expected to be relatable to the electrons entering the conduction band and responsible for the electrical properties. In order to get deeper insight into the role that the different crystal surfaces play in determining the electrical response, the amount of $V_0{}^+$ centers detected after CO/Ar treatment of OCT, DOD, and NBA SnO_2 nanoparticles was related to the structure of the different exposed surfaces and plotted versus the area of exposed {110} crystal face. The very sensitive OCT nanoparticles do not possess any {110} crystal face, while entirely expose the {221} surfaces; OCT shows the highest amount of $V_0{}^+$. A lower amount of paramagnetic defects was observed in DOD nanoparticles which have a certain amount of {110} crystal faces beside the {111} ones. NBA nanoparticles contain high amount of {110} crystal face, while no paramagnetic defects (Fig. 5.43).

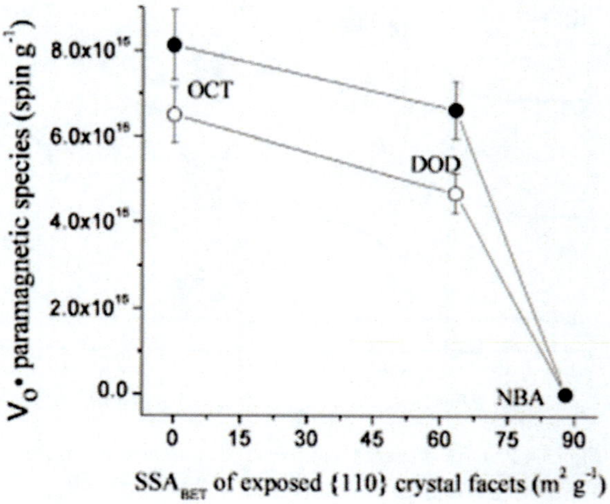

Fig. 5.43 V_0^+ paramagnetic species against exposed {110} crystal face, for the different nanoparticles [26]

5.52.3 Electrical Response

Referring to the defect identification, reported for the differently shaped nanoparticles, Fig. 5.44 reports the corresponding electrical response under CO/argon atmosphere treatment.

The highest response appears for the OCT nanoparticles which also have the highest amount of V_0^+ defects [26].

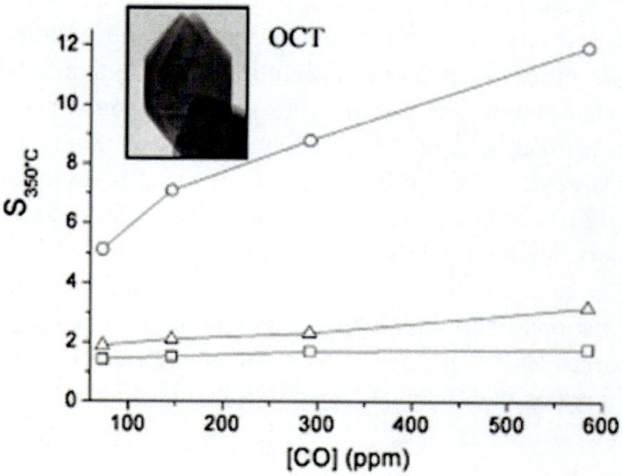

Fig. 5.44 Electrical response for ○ OCT, Δ DOD, □ NBA

5.53 Chemical Message

The results can be associated with the arrangement and coordination of the surface atoms in the different crystal facets exposed by the nanocrystals. High energy {221} and {111} faces, mainly occurring in OCT and DOD, present a higher density of under-coordinated Sn centers compared to the low energy {110}, predominant in NBA. Thus, the first two faces appear more able to stabilize defects. The reactivity which originates the electrical response happens at atomic level.

5.54 The Case of Silica–Natural Rubber Composites

The study of silica–rubber (natural or not) composites is of fundamental importance for designing tires [27, 28]. SiO_2 particles are now becoming fillers alternative to the carbon black, previously used. Fundamental to obtain efficient good reinforce of the rubber is the particle dispersion and as a consequence the prevalence of the filler–rubber interaction over the filler–filler.

The research makes use of the in situ solgel synthesis of filler inside a natural rubber matrix and of particles with different shapes, in order to check the most suitable composite products.

The hydrolysis and condensation leading to filler particles were performed on the following silica precursors.

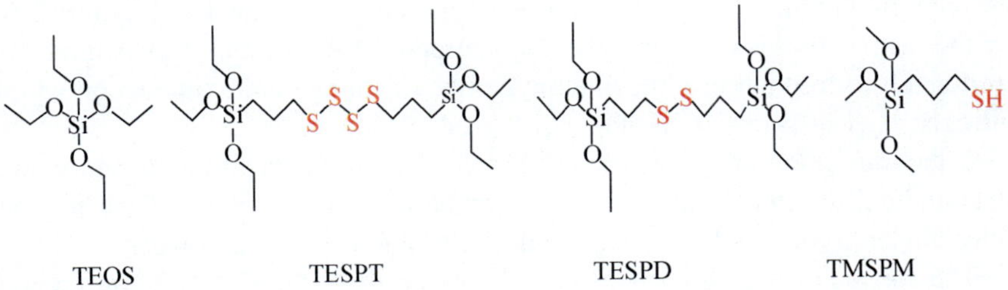

Silica precursors used in sol-gel procedure

All molecules display anisotropic structure and in principle should have different filler–rubber interactions, due to the different interfaces. Only TEOS gives rise to spherical particles. The dispersion is guaranteed by the solgel procedure of synthesis. The aim is to relate the shape of particles with the mechanical properties that means with the filler–rubber interaction.

5.54.1 Morphology of the Dispersed Filler

(NR)-TEOS shows well-assembled aggregates of spherical particles and a homogeneous filler network; (NR)-TESPT and (NR)-TESPD have inhomogeneous shape with both spherical and anisotropic particles, but the aggregates are less compact than in (NR)-TEOS, probably due to a stronger filler–rubber interaction. As for (NR)-TMSPM, large anisotropic particles are visible, while responsible of strong aggregation, this being due to interaction between thiol groups (Fig. 5.45). On the basis of morphological properties, a different rigidity of the rubber composite, due to a different cross-linking extent, is expected. The investigation was performed by the spin probe method.

5.54.2 Spin Probe Procedure

ESR measurements were performed on composites (NR)-TEOS, (NR)-TESPT, (NR)-TESPD, and (NR)-TMSPM and on the reference sample (NR) SiO_2 after introduction of the TEMPOL spin probe in the rubber matrix. TEMPOL (in Fig. 5.46) is a nitroxide radical largely used in the procedures assisted by the ESR spectroscopy. The radical embedded into the matrix allows to detect the motion of the polymer molecules. In fact, the room-temperature ESR spectrum of TEMPOL is sensitive to the dynamics of the nitroxide molecules, which is determined by the rotational correlation time τ_R (time required for a complete radical rotation around its axis). Fast motion occurs when nitroxide has no or very few mobility restrictions by the matrix. In this case, the resulting isotropic ESR spectrum shows three sharp well-resolved resonances with small hyperfine coupling due to interaction with the nitrogen nuclear spin ($I = 1$) (Fig. 5.47).

If the nitroxide mobility is limited by physical or chemical interaction with the surrounding, the line width and the hyperfine coupling increase. If the motion is fully hindered, the ESR anisotropic lines resemble those of a powder.

The spectra of NR composites here discussed show the ESR spectrum reported in Fig. 5.48.

A very strong broad anisotropic signal ($g_1 = 2.023$, $g_2 = 2.005$, $g_3 = 1.981$) associable to spin probe slow motion in a high-density matrix is observed. The resonances overlap to the much less intense narrow isotropic lines due to fast motion spin probe in the free volumes of the matrix. Figure 5.48 reports the hyperfine coupling values of the outer resonances for the two species (iso or fast and aniso or slow motion). From the highness of the iso- and aniso-resonances, it has calculated the ratio I_{ANISO}/I_{ISO} to which the probe motion in the rubber matrix can be associated. Higher value of this ratio means prevalence of slow motion. All

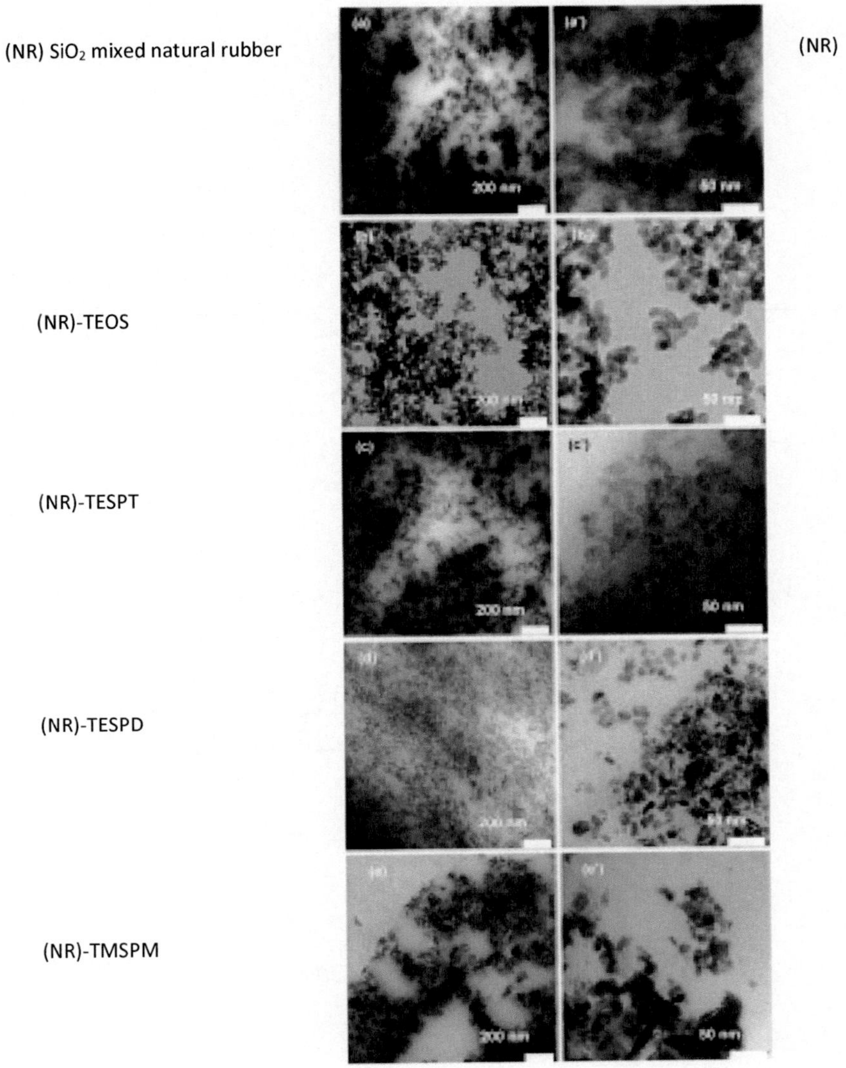

Fig. 5.45 TEM micrograph of the silica nanocomposites obtained from the indicated precursors. (NR) SiO_2 is the reference sample prepared by mechanical mixing. Magnification in the right column [27, 28]

the composites obtained by solgel procedure in the rubber show higher rigidity than the conventional (NR) SiO_2 obtained by mixing, and it appears that the anisotropy of the silica precursor is associated both with the cross-linking of the matrix and with the filler–rubber interaction.

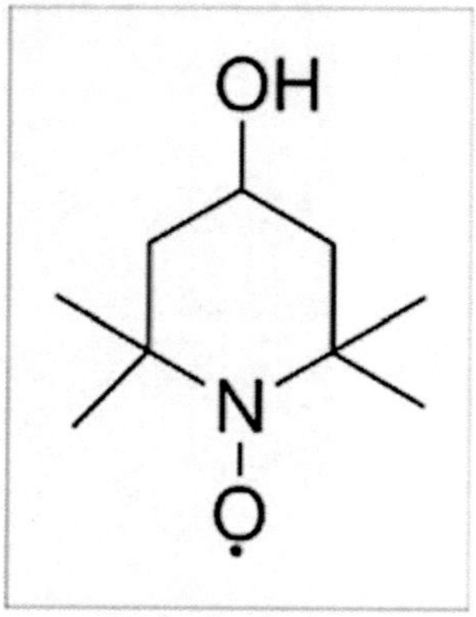

Fig. 5.46 TEMPOL scheme

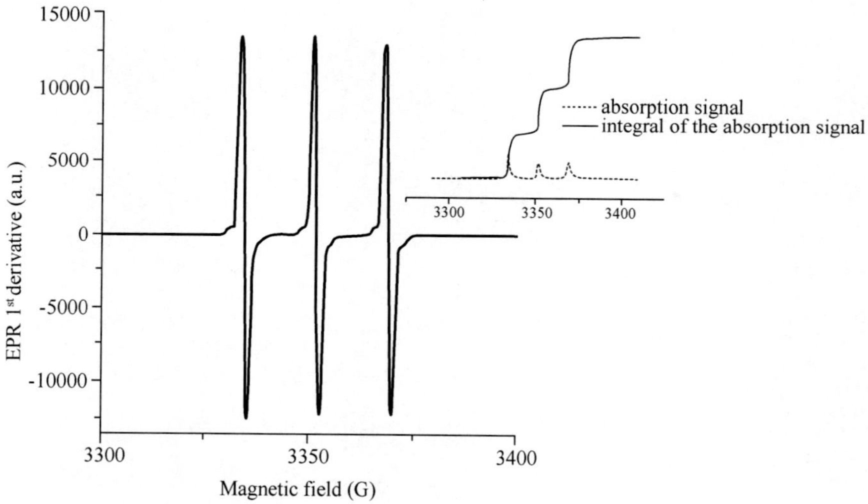

Fig. 5.47 ESR spectrum of TEMPOL

5.55 Chemical Message

The results demonstrate that the shape of the filler particles can have great influence on the mechanical properties of the rubber composites and indicate a pathway to design tires with desired properties. Let us also consider that the excess of mobility of the filler particle also constitutes a loss of energy during the rolling of tire and this opposes the principles of a green material.

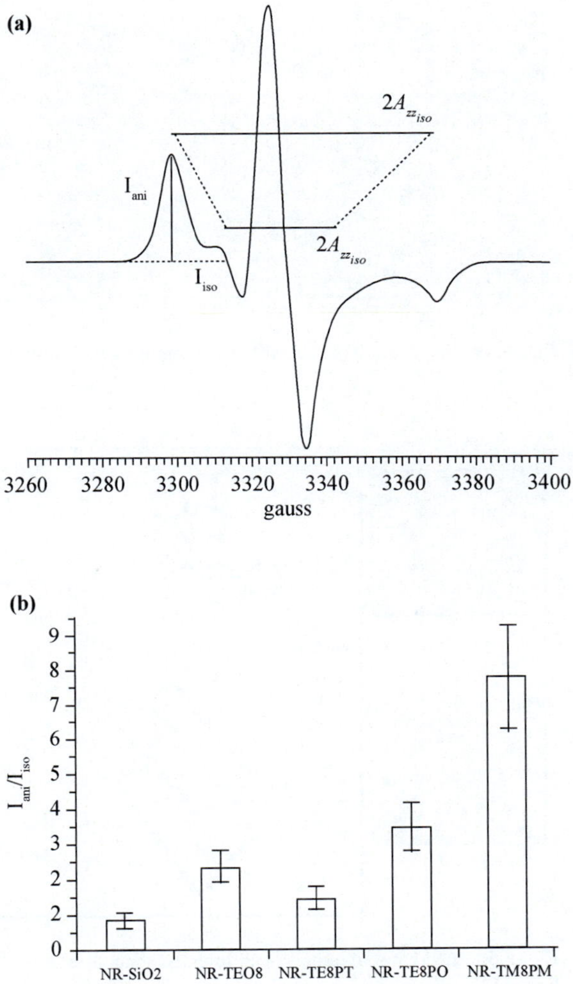

Fig. 5.48 **a** ESR spectrum of (NR)-TMSPM and **b** I_{ANISO}/I_{ISO} of the indicated composites [27, 28]

Beginning from the statement that the filler–rubber interaction increases in anisotropic filler, new composites containing shape-controlled particle with different aspect ratios were prepared (Fig. 5.49).

Atomic force micrographs (Fig. 5.50) detail that the anisotropic particles are surrounded by overlapped higher-density polymer layers which better reinforce the cross-linking of the polymer and enhance the mechanical properties of the composite.

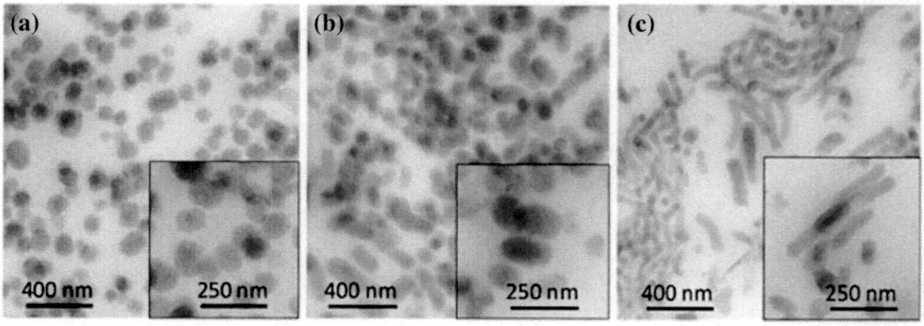

Fig. 5.49 Styrene-butadiene rubber containing anisotropic silica particles [27, 28]

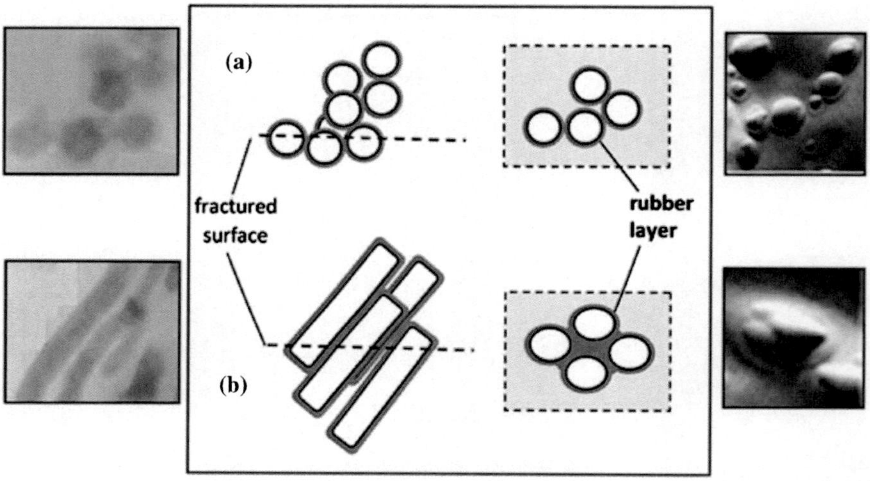

Fig. 5.50 AFM micrographs of differently shaped silica particles and the surrounding polymer layers [27, 28]

5.56 The Case of Cis Diamino Dichloro Platinum (Cisplatin) Antitumor Drug [Pt(NH)$_3$Cl$_2$]: Mechanism of Combinational Chemo-radiotherapy

Cisplatin (CDDP) is one of the most widely used anticancer drugs [29]. Despite its great success, CDDP shows severe toxic side effects and acquired resistance. For these reasons, researchers tried to develop new platinum anticancer drugs, based on the Pt coordination properties. The number of employable new drugs was very low compared with that of the synthesized compounds; thus, it seemed of greater benefit to modulate the pristine CDDP. Which kind of modulation? Dose modulation?

Different ways of administration? Administration combined with radiation therapy? In order to validate a therapeutic protocol, the study of the action mechanism is mandatory, in particular when using combined radiation therapy. The commonly proposed CDDP mechanism suggests that the major damage to DNA is induced after the hydrolysis of the CDDP to cis-$Pt(NH)_3$ and further attaches to DNA bases, without any electron exchange process. In the case of chemo-radio combination therapy, electron transfer processes should be implied and an additional pathway is expected to be active and responsible for the enhanced effects.

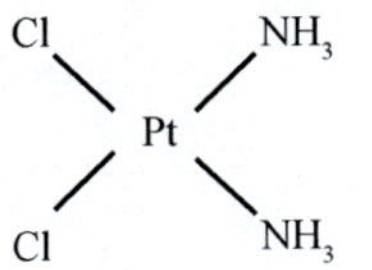

Structure of Cis diamino dichloro platinum (CDDP)

Assuming that the radiation source generates electrons, it may be expected that they go to the metal center, which originally has the Pt(II) d^8 electronic configuration. This can have influence on the coordination geometry of the platinum complex. Electronic spectroscopy can help in detecting the metal center modifications and suggesting an activated form of the drug.

5.57 Discussion of the Case

Figure 5.51 reports the electronic spectrum of cisplatinum and the intensities of the bands, varying the drug amount in solution. Considering the electronic spectra of a simple tetracoordinated planar Pt(II) compounds (Fig. 5.52), the planar geometry can be spectroscopically assigned to cisplatinum. Three absorption bands are attributable to d-d transition $3a_{1g} \rightarrow 3b_{1g}$ $2e_g \rightarrow 3b_{1g}$ $2b_{2g} \rightarrow 3b_{1g}$ that means $d_z^2 \rightarrow d_{x^2-y^2}$ $d_{xz,\ yz} \rightarrow d_{x^2-y^2}$ $d_{xy} \rightarrow d_{x^2-y^2}$, with two transitions very similar in energy (see the energy trend of $PtCl_4^{2-}$).

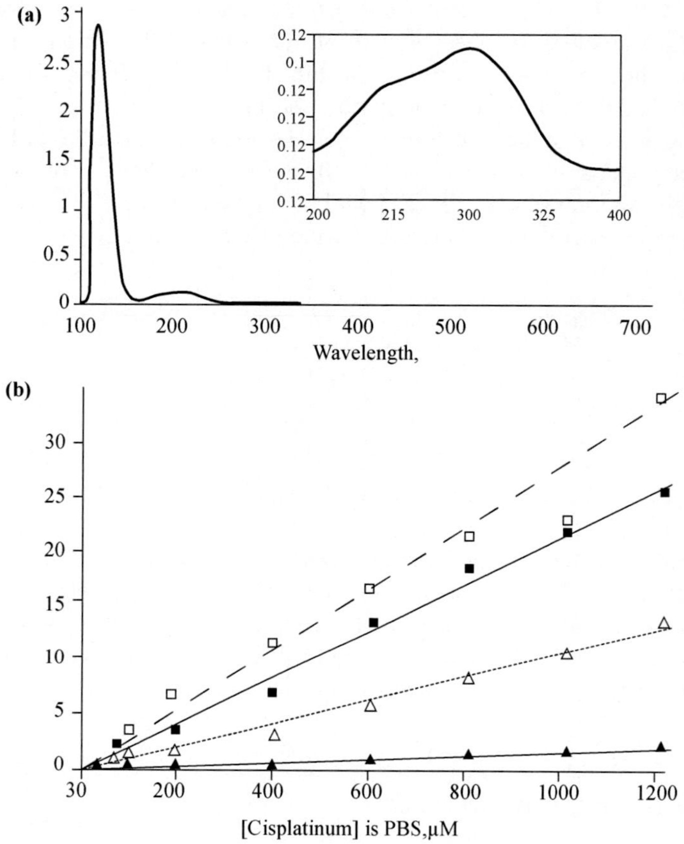

Fig. 5.51 **a** Electronic spectrum of cisplatinum in the UV–visible region. **b** Integrated intensity of the bands as a function of the drug amount [29]

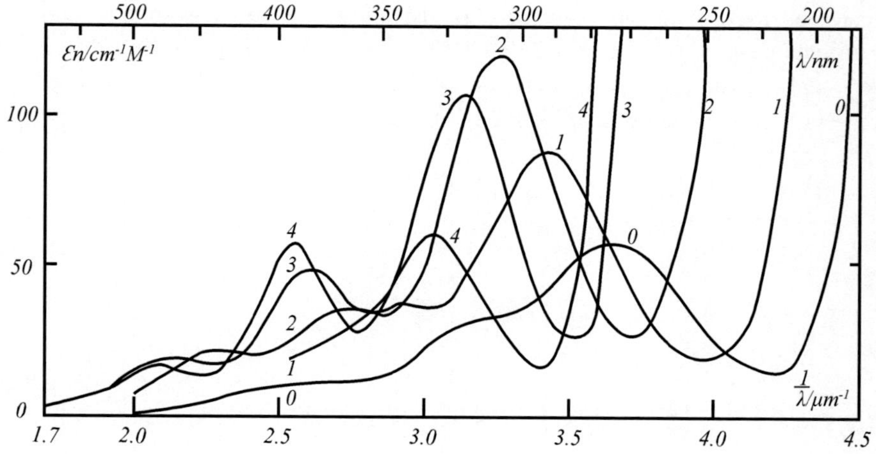

Fig. 5.52 d-d spectra for $[PtCl_n (H_2O)_{4-n}]^{2-n}$ (n = 0, 1, 2, 3, 4); the spectrum for n = 2 is a cis–trans mixture

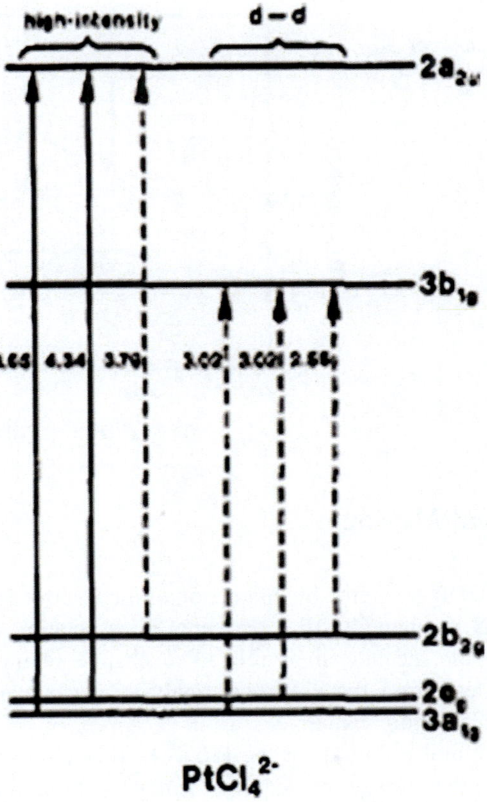

Energy level trends of $PtCl_4^{2-}$

It is expected that any distortion from the planar symmetry D_{4h} or change in the Pt electronic configuration can be revealed by the spectral analysis. In the literature, it has been reported that excitation of CDDP by 318 nm irradiation induces variations in the absorption electronic spectrum, specifically the bands at about 300 nm loss intensity (Fig. 5.53).

A possible interpretation is that the excitation at 318 nm reduces Pt(II) centers to Pt(I) consequently destroying the complex symmetry. A possible reaction pathway could be

$$e^- + Pt(NH_3)_2Cl_2 \rightarrow [Pt(NH_3)_2Cl_2]^- \rightarrow Pt(NH_3)_2Cl^{\cdot} + Cl^-$$
$$e^- + Pt(NH_3)_2Cl \rightarrow [Pt(NH_3)_2Cl]^- \rightarrow Pt(NH_3)_2I^{\cdot} + Cl^-$$

where the formation of the Pt radical, with less steric hindrance than CDDP, is triggered by the Pt reduction. The radicals, well penetrating, are responsible for the damage to DNA of human cells. It could be argued that higher the acidity of Pt(II) centers, modulated by the ligands, higher the attitude of Pt to be reduced.

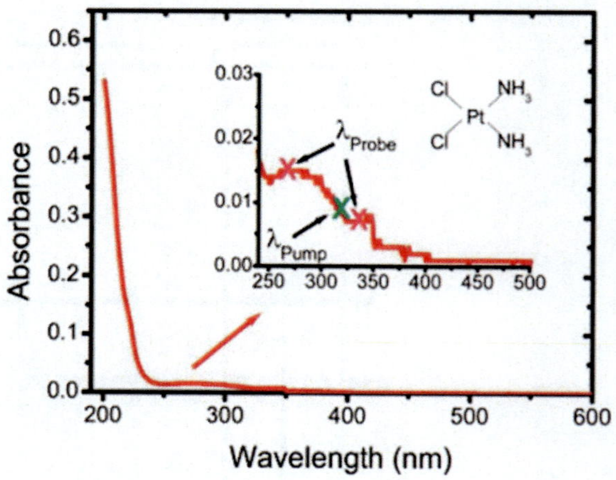

Fig. 5.53 Electronic spectrum of CDDP in water irradiated at 318 nm [29]

5.58 Chemical Message

The radicals of CDDP generated by interaction of the electrons with the drug trigger the damage to DNA, when CDDP is delivered in association with radiotherapy. It seems to us that the mechanism is related to change of molecular structure of CDDP, well evident from the electronic analysis, consequent to the change of oxidation state of platinum center.

Given the structural proposal for the Pt/DNA adducts, responsible for the permanent damage to the biomacromolecule, it is expected that the Pt(I) center, inside $Pt(NH_3)_2Cl^{\cdot}$ and $Pt(NH_3)_2I^{\cdot}$, is reoxidized to Pt(II) after the interaction with DNA and it releases one electron. This electron can further interact with CDDP molecules as in a true catalytic process (Fig. 5.54).

The crucial event seems the formation of radicals triggered by Pt reduction; this process is expected to be favored by the attitude of P(II) to accept electrons. The proposal of an eventual CDDP analogue could be based on the coordination of Pt (II) with electron accepting ligands. One example is the (R,R)-1,2-diaminocicloesano(etandioato-O,O)platino, known as oxaliplatin (Scheme in Fig. 5.55).

CDDP and its homologues have severe side effect during the therapy, when used as drug alone. In the case of combined chemo-radiotherapy, the dose of Pt drug could be decreased as the radicalic and catalytic mechanism implies chain reactions with the DNA.

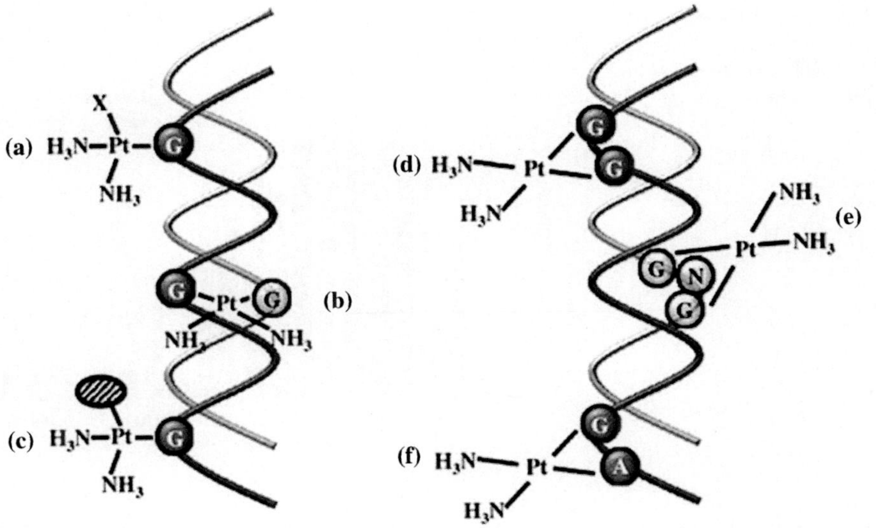

Fig. 5.54 Cisplatin DNA adducts: **a** cisplatin bound monofunctionally to guanine (X—original chloride or a hydroxyl group); **b** interstrand cross-link; **c** cisplatin guanine—protein cross-link; **d** GpG intrastrand cross-link; **e** GpNpG intrastrand cross-link (N represents a base); **f** ApG intrastrand cross-link

Fig. 5.55 Scheme of oxaliplatin

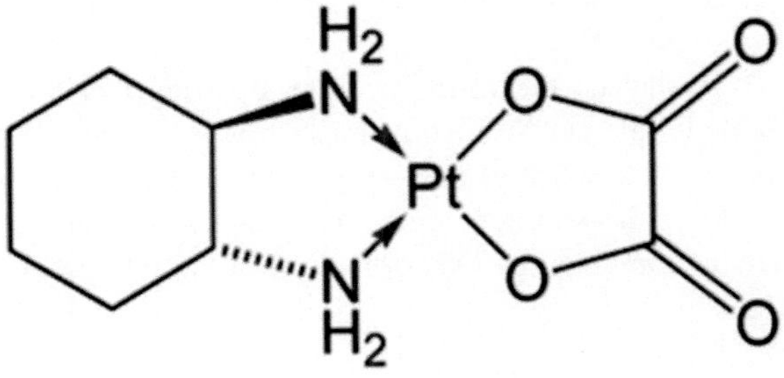

5.59 The Case of Graphite Particles that Induce Reactive Oxygen Species in Cells

It is commonly accepted that the toxicity of carbonaceous particulate matter (PM) is related to the production of reactive oxygen species (ROS) which induce damage in the exposed cells [30]. It is also known that PM produced during the combustion processes consists of a carbonaceous core "dressed" with organic and/or inorganic materials. In spite of the knowledge, the role of these materials in the production of ROS remains unclear. The research aims at understanding whether "naked" carbonaceous particles alone are able to produce ROS in cell systems, even in the

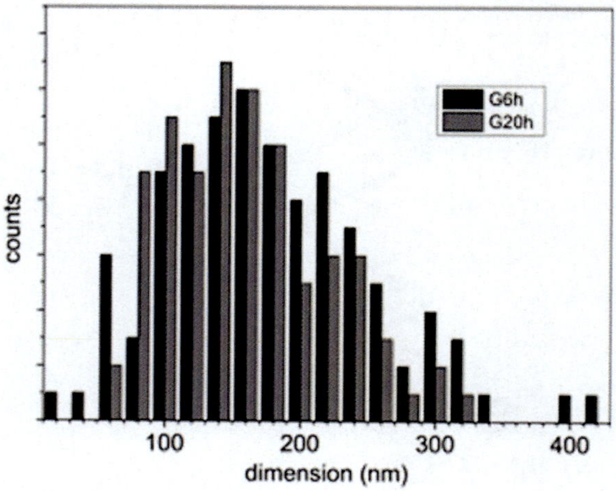

Fig. 5.56 Size distribution of the graphite particle milled for 6 h (G6h) and 20 h (G20h) [30]

absence of the organics or the metal particles that usually catalyze the ROS formation. Object of investigation has been graphite particles of nanometric dimension, obtained by milling micrometric graphite for different times (6 or 20 h). Figure 5.56 reports the size distribution of the particle size.

The large planes of benzene rings in graphite are never perfect as they contain vacancies, dislocations, and carbon atoms in sp^3 or sp hybridization that generate structural and chemical defects. Moreover, graphite flakes have irregular boundaries which produce additional defects like hybridized sp^2 carbons having different configurations.

Basing on the wide literature reporting that frequently the defects in solids are able to interact with oxygen and to reduce it to superoxide O_2^-, the purpose was to investigate whether the oxygen is reduced by graphite defects and if this happens both in the absence and in the presence of human cells and then to compare the reduction extent (ROS formation) in the two cases.

5.60 Discussion of the Case

The experiments, mainly based on electron spin resonance measurements, were performed on G20h nanoparticles in aqueous medium, in the presence of the reducing agent dithiothreitol (DTT) (Fig. 5.57), which triggers the reduction of oxygen.

Subsequently, the spin trap method has been employed to evidence the formation of the ROS species. The spin trap procedure consists in the detection of short-living ROS, namely O_2^- or $OH^{\cdot}$, that become visible only when they form long-living radical adducts after reaction with spin trap molecules. One of the most common spin trap molecules is DMPO (Fig. 5.58).

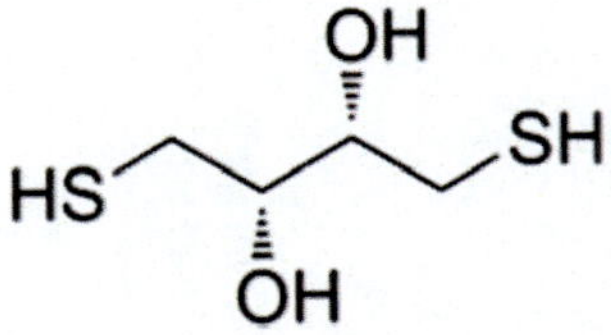

Fig. 5.57 DTT molecular structure

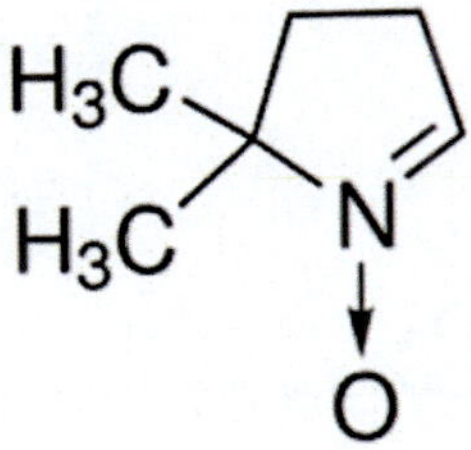

Fig. 5.58 Molecular structure of 5,5-dimethyl-1 pyrroline N oxide (DMPO)

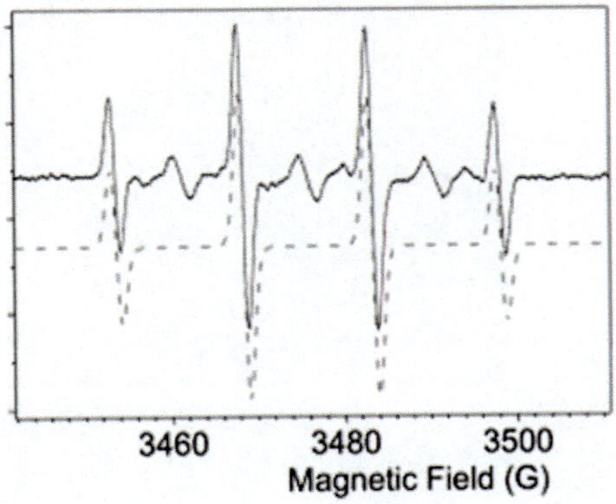

Fig. 5.59 ESR spectrum of spin adduct in cell-free systems, after addition of DMPO and DTT, reacting 20 min [30]

Figure 5.59 shows the ESR spectrum of an aqueous solution containing G20 and DTT after 20-min reaction time. The spectrum of DMPO-OH adduct was observed, deriving from the reaction of OH˙ radical with DMPO. The four major peaks arise from the interaction of the unpaired electron with N ad H atoms and have been reproduced (dashed lines in Fig. 5.60) by $\mathbf{A_N}$ 14.87 G and $\mathbf{A_H}$ 14.81 G hyperfine constant values.

At the reaction time of Fig. 5.60, the adduct DMPO-OOH formed by interaction of the O_2^- radical with DMPO decomposes and it has been demonstrated that DMPO-OH adduct arises from the DMPO-OOH decomposition. The rate of radical formation increases in the presence of graphite particles (Fig. 5.60).

Given that the presence of carbon particles catalyzes the ROS formation, the system has been studied in the presence of human lung cells. Graphite particles are well visible inside the cells 2 h exposed (Fig. 5.61).

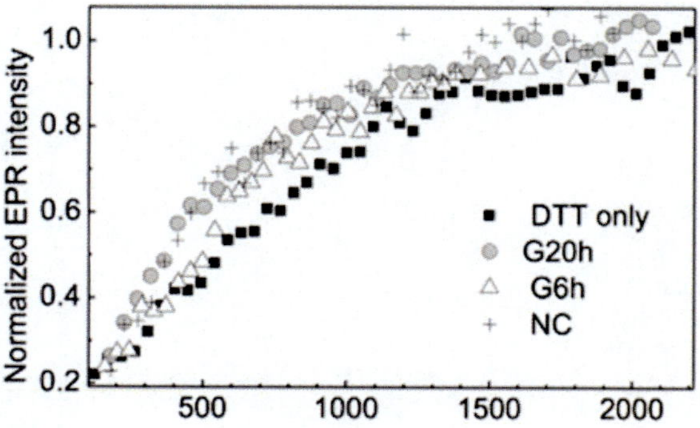

Fig. 5.60 Kinetic of formation of DMPO-OH adduct in the presence of carbon nanoparticles. NC means nanoparticles [30]

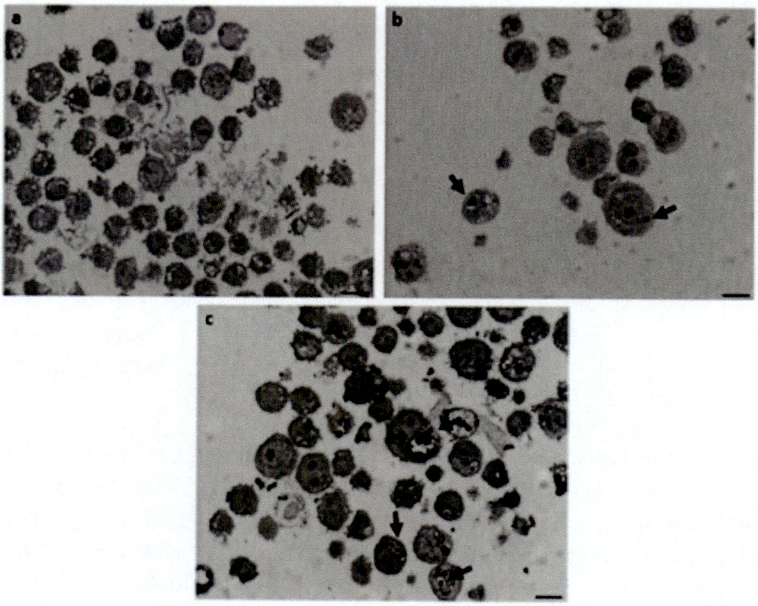

Fig. 5.61 G20h particle uptake as black points inside cells (optic microscope) [30]

A spin trap experiment, parallel to that in the absence of cells, was performed in the presence of them (Fig. 5.62), assuming that cells take the place of DTT as reducing agents.

A careful fit of the ESR spectra, taking account of all species, shows a light increase of the ROS species in the presence of graphite.

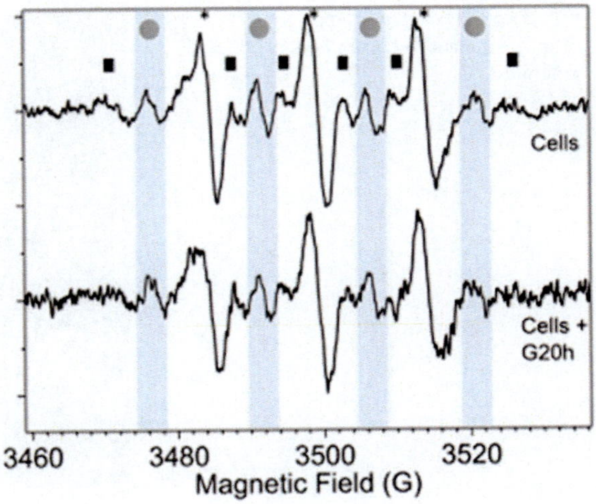

Fig. 5.62 ESR spectra of reaction of DMPO in cells in the absence and in the presence of G20h particles. The light blue shadow areas highlight the signals associated with DMPO.OH adduct [30]

5.61 Chemical Message

The presence of organic or inorganic catalysts on the surface of PM is not necessary to produce ROS. They derive from the defects located at the graphite surface; thus in a number of cases, also unexpected, the reduction of oxygen to radical is active. The toxicity over long-term exposure remains a question to be investigated; however, it has to be considered that ROS production may occur over a long time, with heavy consequences by toxicity.

5.62 The Interaction of Graphite Defects with Oxygen Treated by Structural Studies

In the previous paragraph, it has been shown that the reduction of oxygen is promoted by graphite and specifically the graphite surface defects are responsible for activating the process [31]. In general, the study of a catalytic process implies both the evaluation of the products and the detection of the catalytic intermediates. This is the most difficult goal, as the structure of defects has to be investigated as well as their attitude to transfer electrons. It is well known that the structure of nanosized graphite particles is rather defective as the milling procedure introduces different types of defects. It reduces the size of the graphite sheet from 160 to 9 nm (Fig. 5.63), partly destroying the electronic delocalization, and changes the C hybridization from sp^2 to sp^3. Moreover, the milling promotes the formation of

Fig. 5.63 SEM micrographs of **a** pristine and **b** ball-milled graphite [31]

zigzag and armchair edges and of spin localized non-bonding π electron states in the zigzag periphery of nanosheets. The catalytic action of nanosized graphite is attributed to the presence of non-bonding π electron states in the zigzag periphery of nanosheets which appear to be implied in the electron transfer from the reductant C to oxygen.

5.63 Discussion of the Case

5.63.1 Electrochemistry

Electrochemical measurements allow to associate the current signal with the variation of the reduction potential. Using as electrodes *milled graphite* (BM), *pristine graphite* (PG), and *glassy carbon* (GC) (support for comparison), it resulted in increase of the current signal in BM and PG with respect to GC, due to the surface area. This is associated with an increase of reduction potential (Fig. 5.64).

It appears that the electron charge transfer from the carbon electrode to the oxygen molecule occurred at lower energy in the presence of defective material (BM ball-milled graphite) ($\Delta E = 0.13$ eV) [31].

5.63.2 Electron Spin Resonance Measurements

This technique allows to connect the thermodynamic–electrochemical data, having macroscopic character, to the more important structural data. Figure 5.65 reports the ESR spectra of a different defective graphite. Under high vacuum, the pristine graphite showed weak resonance lines centered at g = 2.007–2.008. No significant changes were observed contacting with O_2 and returning to inert atmosphere. The signal has been assigned to delocalized π electrons on the graphite sheets.

After milling, a sharp symmetric and intense resonance appears at g = 2.0038, this signal being assignable to localized electrons on peripheral C centers. In fact, it is well known that the nanographite obtained by ball milling contains localized spin of non-bonding π electron states (edge states) in the zigzag periphery of graphite sheets. This ESR signal is very sensitive to the presence of oxygen and shows a very strong decrease by oxygen contact. Thus, the edge states of BM interact with oxygen giving an evident spin coupling.

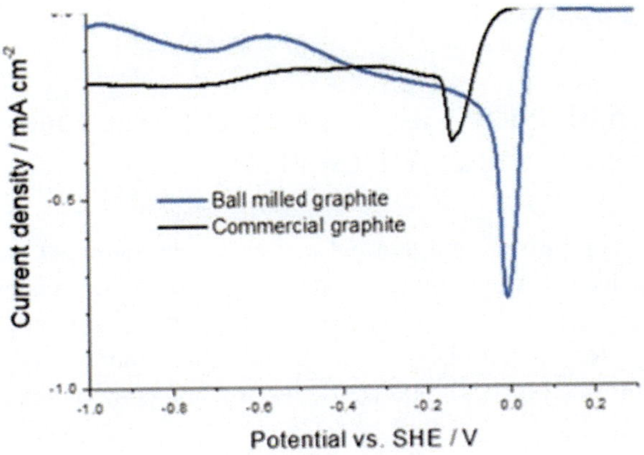

Fig. 5.64 Plot of the current density versus reduction potential for the indicated materials

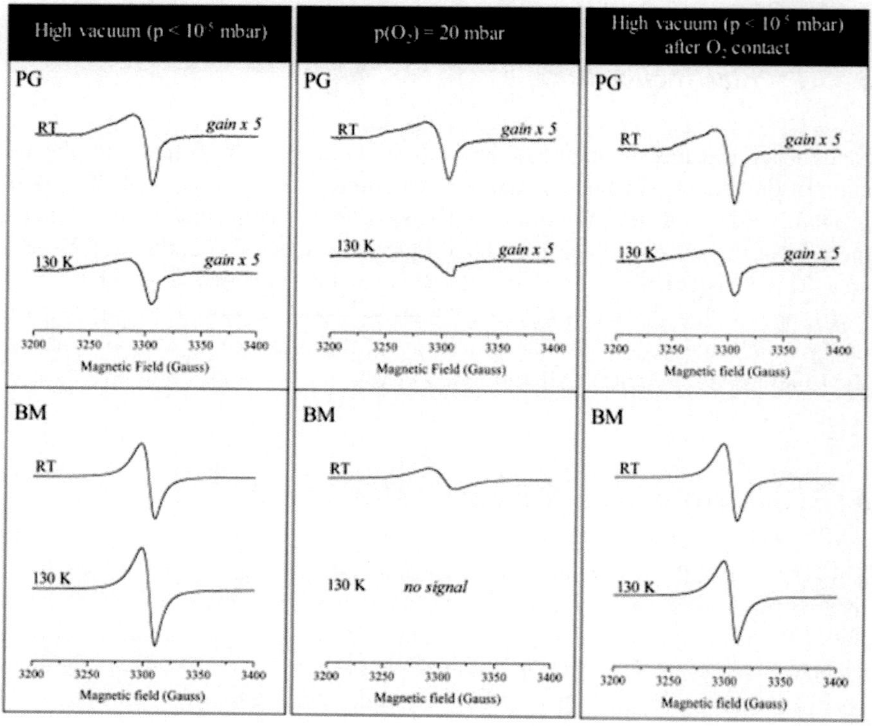

Fig. 5.65 ESR spectra of the indicated samples, under the reported experimental conditions [31]

5.64 Chemical Message

The structure defects of graphite play a determinant catalytic role in the oxygen reduction. They in fact govern the activation energy of the transformation of O_2 to its reduced forms O_2^- O_2^{2-} O^{2-}. Several catalytic processes may take profit from this result, both chemical and biochemical.

5.65 The Case of Persistent Free Radicals in the Atmosphere

Combustion and thermal processes are dominant sources of air pollution and produce toxic products, specially particulate matter (PM), polycyclic aromatic hydrocarbons, and environmentally persistent free radicals (EPFRs) [32]. These later generate DNA damage and induce pulmonary dysfunction. EPFRs on PM are formed through interaction of aromatic compounds with transition metal oxides via

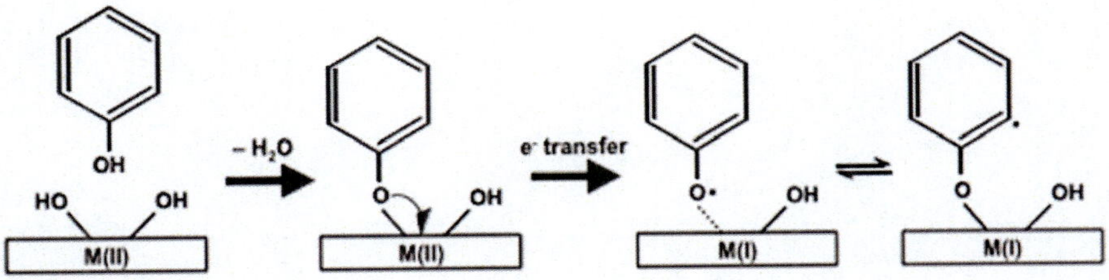

Fig. 5.66 Mechanism of EPFR formation on inorganic oxides

surface-mediated processes. This results in the formation of surface-bound radical species, which persist in the atmospheric environment for days and generate in water the reactive oxygen species (ROS). The biological unbalance between the ROS ($OH^{\cdot}$, O_2^{-}, H_2O_2, etc.) and the molecules that control the cellular homeostasis leads to oxidative stress associated with respiratory and cardiovascular diseases. Thus, EPFRs can be thought the major responsibility for the biological damage.

Scheme (Fig. 5.66) reports the most accepted general mechanism by which the radicals are formed through interaction of the organic molecule with the surface metal centers of a metal oxide (e.g., CuO). As an example (Scheme A) in the radicalization of phenol catalyzed by CuO, the organic molecule interacts with the oxide and transfers electrons to the metal center and the reduction gives stable radicals.

In spite of these suggestions, very few investigations about the mechanism are concerning with the transition metal electronic state. Moreover, no attention has been deserved to the interaction with benzene, one of the most diffuse pollutant agents.

5.66 Discussion of the Case

The analysis begins from the computational approach aimed at rationalizing the formation of EPFRs following the benzene interaction with CuO. The suggested mechanism is shown in Fig. 5.67. Two pathways depart from a *side* and an *edge* O_2 *on* bonding to Cu(II) centers, A_1 and B_1, respectively, which become equivalent after interaction with benzene. In the step (A_4), a hydrogen transfer takes place yielding the stable A_5 intermediate. For this species, both a triplet state and a biradical singlet state (antiferromagnetic) can be suggested without change of electronic state in copper.

Basing on these results, a real system CuO/SiO_2 was studied, having care that the catalyst components were well dispersed and in intimate contact (Fig. 5.68).

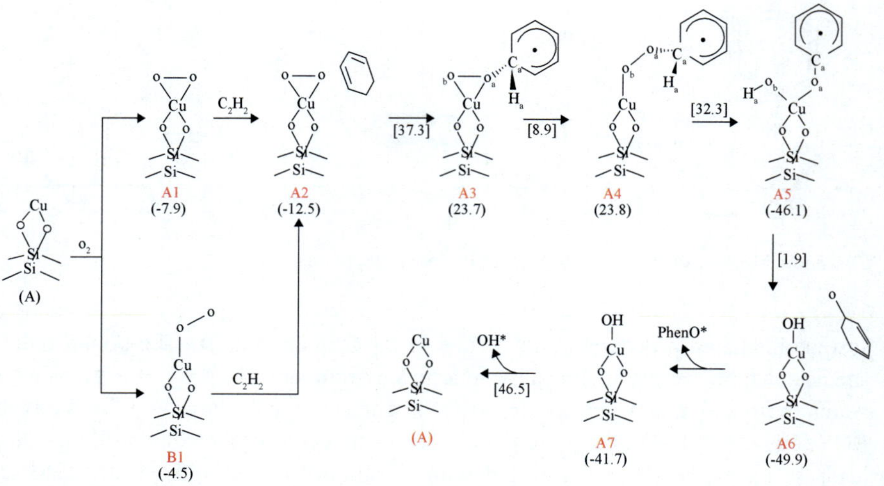

Fig. 5.67 Mechanism of benzene hydroxylation. In the figure, the reaction scheme and the species with the relative electron energies of intermediates with respect to separate reagents (round parentheses) and the activation energies (square parentheses) [32]

The ESR investigation carried out on copper centers of CuO/SiO_2 before interaction with benzene gave the spectra in Fig. 5.69.

It is consistent with a d^9 configuration attributable to Cu(II) centers. The magnetic parameters are well defined $g_{\parallel} = 2.384$, $g_{\dagger} = 2.071$, $A_{\parallel} = 128$ G and suggest an octahedral distorted symmetry.

These results agree with that from the diffuse reflectance spectrum in Fig. 5.70, where the electronic spectrum, recorded before the benzene absorption, shows the lines of a distorted octahedral Cu(II) center.

The contact with benzene caused modification in the catalyst ESR spectrum (Fig. 5.71).

In fact, the Cu(II) resonances decrease in intensity after contact with benzene, while arise a new sharp signal at $g = 2.003$, stable for 24 h. This later can be assigned to phenoxy radical. No hyperfine interaction with copper nucleus was observed suggesting that the radical is stabilized by interaction with the oxide, not with copper centers. The decrease in Cu(II) intensity signals can be associated with the interaction of the phenoxy radical with the d^9 center, which leads to antiferromagnetic coupling.

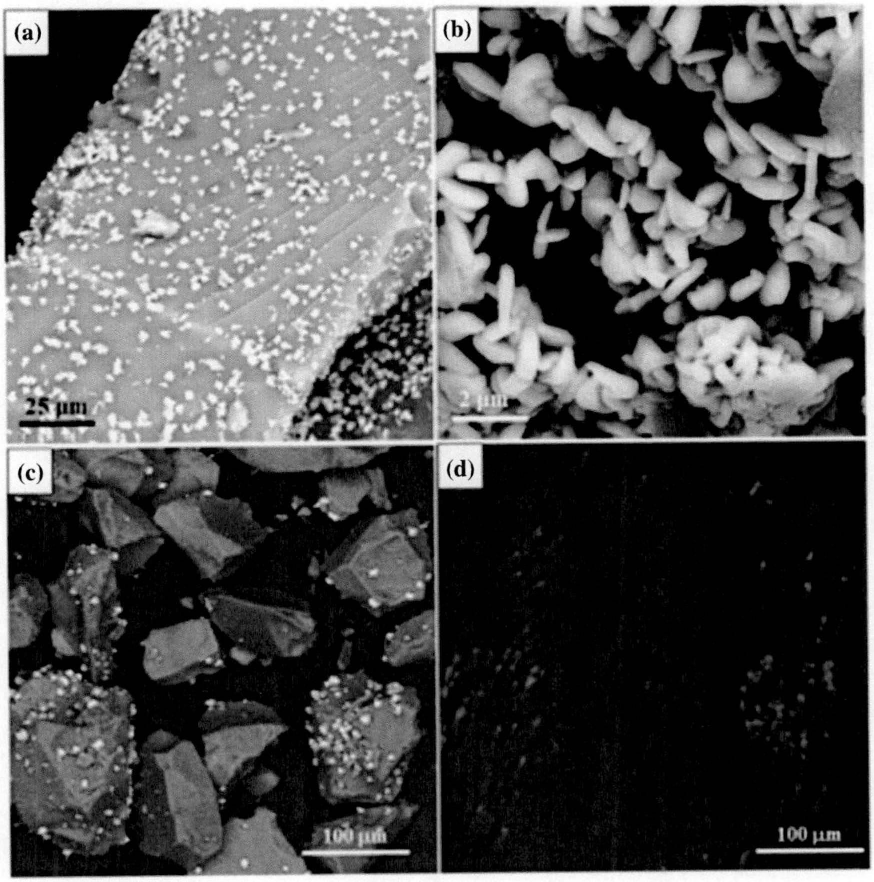

Fig. 5.68 **a–c** SEM micrographs of CuO/SiO_2, **d** elemental mapping of Cu [32]

5.67 Chemical Message

In the presence of O_2, as it is in the atmosphere, even aromatic hydrocarbons can generate persistent radical species. This happens when the transition metal ions adsorb molecular oxygen which act as bridge with the hydrocarbon itself. The driving force is constituted by the oxygen adsorption, and no electron transfer is detected.

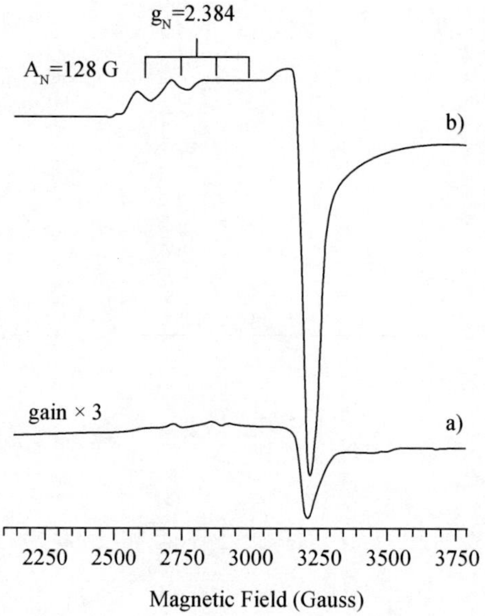

Fig. 5.69 ESR spectrum of CuO/SiO_2 recorded at **a** 298 and **b** 130 K [32]

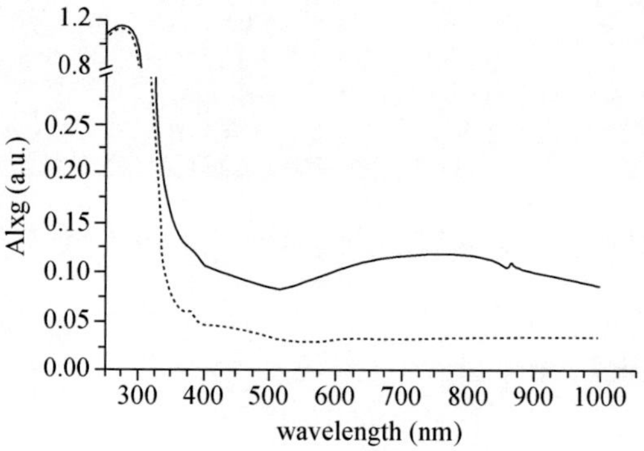

Fig. 5.70 Diffuse reflectance spectrum of CuO/SiO_2 (continuous line) and SiO_2 (dashed line) [32]

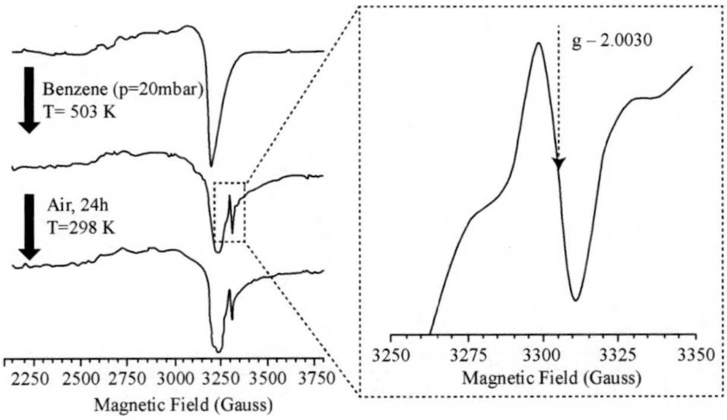

Fig. 5.71 ESR resonances of CuO/SiO_2 after contact with benzene [32]

General remarks

The case studies described in this chapter demonstrate that, basing on the symmetry arguments, the electronic properties of molecules, clusters, surfaces, and also particles can be conveniently described. Being the electronic properties responsible for the chemical functionality, the symmetry can be definitely related to the functionality of all the species.

Thus, the most simple and visible property, the shape, is able to suggest possible reactions and applications of the chemical units, being an important bridge between the base and the applied science.

Appendix

Tables of Characters Used in the Reported Molecules are in Appendix

T_d	E	$8C_3$	$6C_2$	$6S_4$	$6\sigma_d$		
A_1	1	1	1	1	1		$x^2+y^2+z^2$
A_2	1	1	1	−1	−1		
E	2	−1	2	0	0		$(2z^2-x^2-y^2; x^2-y^2)$
T_1	3	0	−1	1	−1	$(R_x; R_y; R_z)$	
T_2	3	0	−1	−1	1	$(x; y; z)$	$(xy; xz; yz)$

O	E	$6C_4$	$3C_2$ $(=C_4^2)$	$8C_3$	$6C_2$		
A_1	1	1	1	1	1		$x^2+y^2+z^2$
A_2	1	−1	1	1	−1		
E	2	0	2	−1	0		$(2z^2-x^2-y^2; x^2-y^2)$
T_1	3	1	−1	0	−1	$(R_x; R_y; R_z)$; $(x; y; z)$	
T_2	3	−1	−1	0	1		$(xy; xz; yz)$

F. Morazzoni, *Symmetry in Inorganic and Coordination Compounds*, Lecture Notes in Chemistry 106, https://doi.org/10.1007/978-3-030-62996-0

O_h	E	$8C_3$	$6C_2$	$6C_4$	$3C_2$ $(=C_4^2)$	i	$6S_4$	$8S_6$	$3\sigma_k$	$6\sigma_d$		
A_{1g}	1	1	1	1	1	1	1	1	1	1		$x^2+(2z^2x^2-(xz,$
A_{2g}	1	1	−1	−1	1	1	−1	1	1	−1		
E_g	2	−1	0	0	2	2	0	−1	2	0		
T_{1g}	3	0	−1	1	−1	3	1	0	−1	−1	(R_x, R_y, R_z)	
T_{2g}	3	0	1	−1	−1	3	−1	0	−1	1		
A_{1u}	1	1	1	1	1	−1	−1	−1	−1	−1		
A_{2u}	1	1	−1	−1	1	−1	1	−1	−1	1		
E_u	2	−1	0	0	2	−2	0	1	−2	0		
T_{1u}	3	0	−1	1	−1	−3	−1	0	1	1	(x, y, z)	
T_{2u}	3	0	1	−1	−1	−3	1	0	1	−1		

D_{2h}	E	$C_2(z)$	$C_2(y)$	$C_2(x)$	i	$\sigma(xy)$	$\sigma(xz)$	$\sigma(yz)$		
A_g	1	1	1	1	1	1	1	1		x^2, y^2, z
B_{1g}	1	1	−1	−1	1	1	−1	−1	R_z	xy
B_{2g}	1	−1	1	−1	1	−1	1	−1	R_y	xz
B_{3g}	1	−1	−1	1	1	−1	−1	1	R_x	yz
A_u	1	1	1	1	−1	−1	−1	−1		
B_{1u}	1	1	−1	−1	−1	−1	1	1	z	
B_{2u}	1	−1	1	−1	−1	1	−1	1	y	
B_{3u}	1	−1	−1	1	−1	1	1	−1	x	

D_{3h}	E	$2C_3$	$3C_2$	σ_h	$2S_3$	$3\sigma_v$		
A_1'	1	1	1	1	1	1		x^2+y^2, z^2
A_2'	1	1	−1	1	1	−1	R_z	
E'	2	−1	0	2	−1	0	(x, y)	(x^2-y^2, xy)
A_1''	1	1	1	−1	−1	−1		
A_2''	1	1	−1	−1	−1	1	z	
E''	2	−1	0	−2	1	0	(R_x, R_y)	(xz, yz)

D_{4h}[a]	E	$2C_4$	C_2	$2C_2'$	$2C_2''$	i	$2S_4$	σ_h	$2\sigma_v$	$2\sigma_d$		
A_{1g}	1	1	1	1	1	1	1	1	1	1		$x^2+y^2,$
A_{2g}	1	1	1	−1	−1	1	1	1	−1	−1	R_z	
B_{1g}	1	−1	1	1	−1	1	−1	1	1	−1		x^2-y^2
B_{2g}	1	−1	1	−1	1	1	−1	1	−1	1		xy

(continued)

(continued)

D_{4h}[a]	E	$2C_4$	C_2	$2C_2'$	$2C_2''$	i	$2S_4$	σ_h	$2\sigma_v$	$2\sigma_d$		
E_g	2	0	−2	0	0	2	0	−2	0	0	(R_x, R_y)	(xz, yz)
A_{1u}	1	1	1	1	1	−1	−1	−1	−1	−1		
A_{2u}	1	1	1	−1	−1	−1	−1	−1	1	1	z	
B_{1u}	1	−1	1	1	−1	−1	1	−1	−1	1		
B_{2u}	1	−1	1	−1	1	−1	1	−1	1	−1		
E_u	2	0	−2	0	0	−2	0	2	0	0	(x, y)	

[a] σ_v passes through the atoms, and σ_d bisects the bond angles

$C_{\infty v}$	E	$2C_\infty^\Phi$	...	$\infty\sigma_v$		
$A_1 \equiv \Sigma^+$	1	1	...	1	z	x^2+y^2, z^2
$A_2 \equiv \Sigma^-$	1	1	...	−1	R_z	
$E_1 \equiv \Pi$	2	$2\cos\Phi$	...	0	$(x, y); (R_x, R_y)$	(xz, yz)
$E_2 \equiv \Delta$	2	$2\cos 2\Phi$	...	0		(x^2-y^2, xy)
$E_3 \equiv \Phi$	2	$2\cos 3\Phi$	...	0		
...	...	...	...	...		

$D_{\infty h}$	E	$2C_\infty^\Phi$	...	$\infty\sigma_v$	i	$2S_\infty^\Phi$	...	∞C_2		
Σ_g^+	1	1	...	1	1	1	...	1		$x^2+y^2,$
Σ_g^-	1	1	...	−1	1	1	...	−1	R_z	
Π_g	2	$2\cos\Phi$	...	0	2	$-2\cos\Phi$	...	0	(R_x, R_y)	(xz, yz)
Δ_g	2	$2\cos 2\Phi$	...	0	2	$2\cos 2\Phi$	...	0		$(x^2-y^2,$
...	...	...	...	...	...	...	...	...		
Σ_u^+	1	1	...	1	−1	−1	...	−1	z	
Σ_u^-	1	1	...	−1	−1	−1	...	1		
Π_u	2	$2\cos\Phi$	...	0	−2	$2\cos\Phi$	...	0	(x, y)	
Δ_u	2	$2\cos 2\Phi$	...	0	−2	$-2\cos 2\Phi$	...	0		
...	...	...	...	...	...	...	...	...		

C_{2v}[a]	E	C_2	$\sigma_v(xz)$	$\sigma_v'(yz)$		
A_1	1	1	1	1	z	x^2, y^2, z^2
A_2	1	1	−1	−1	R_z	xy
B_1	1	−1	1	−1	x, R_y	xz
B_2	1	−1	−1	1	y, R_x	yz

[a] For a planar molecule, the x-axis is taken perpendicular to the plane

spectroscopic states having the same symmetry and the same spin multiplicity do not cross over. In the following are reported the remaining electronic configurations. Consider that the diagrams are valid for both O_h and D_{4h} symmetries, provided the energy trend inversion.

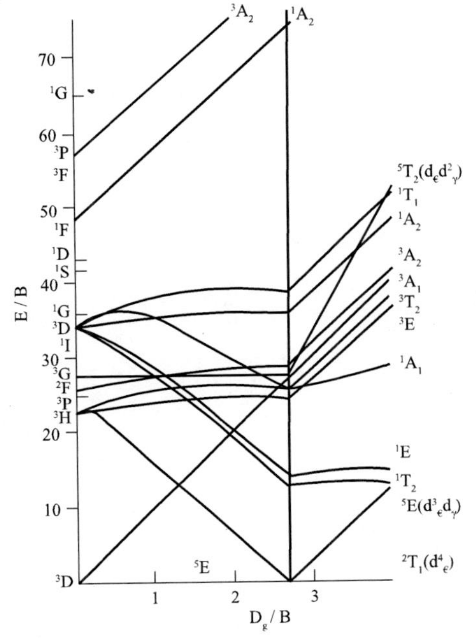

Energy diagram for the configureation d^4

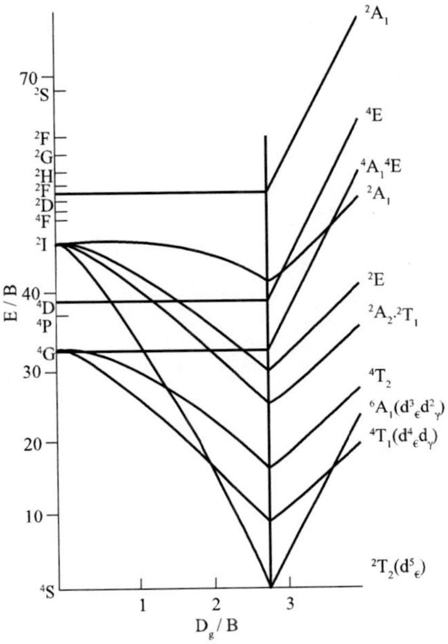

Energy diagram for the configureation d^5

These diagrams show a vertical line which separates two regions of the electronic states, that having the maximum spin multiplicity (high spin) and that the minimum one (low spin). The high spin region is at lower *Dq*.

No energy meaning is attributable to the vertical line.

$\mathbf{d}_\gamma$ means $\mathbf{e_g}$ orbital, $\mathbf{d}_\varepsilon$ means $\mathbf{t_{2g}}$ orbitals. The other symbols are those conventional.

The electronic configurations range from d^2 to d^8; d^1 and d^9 are very simple and constituted by only two right lines, those of T_2 and E states.

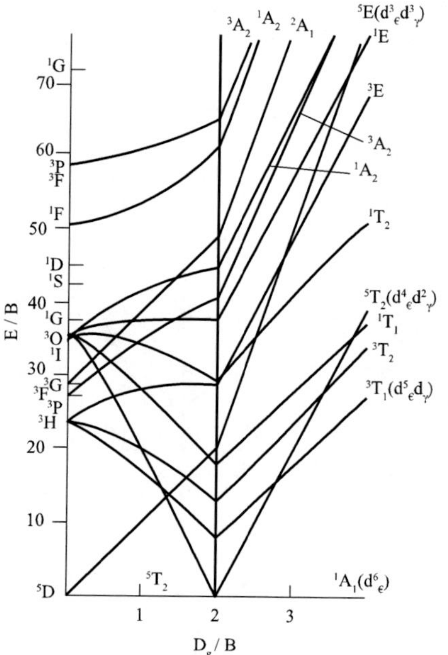

Energy diagram for the configureation d^6

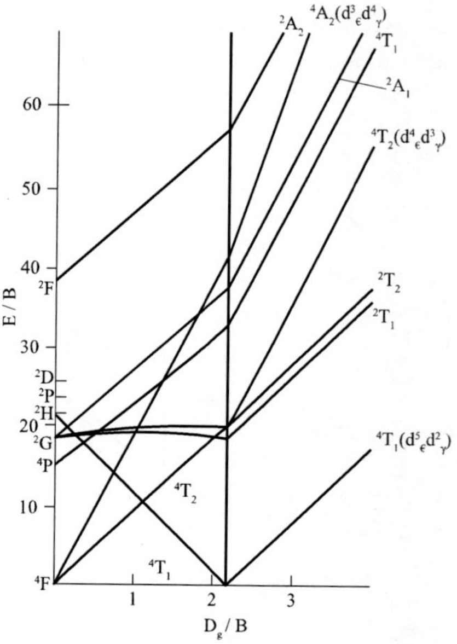

Energy diagram for the configureation d^7

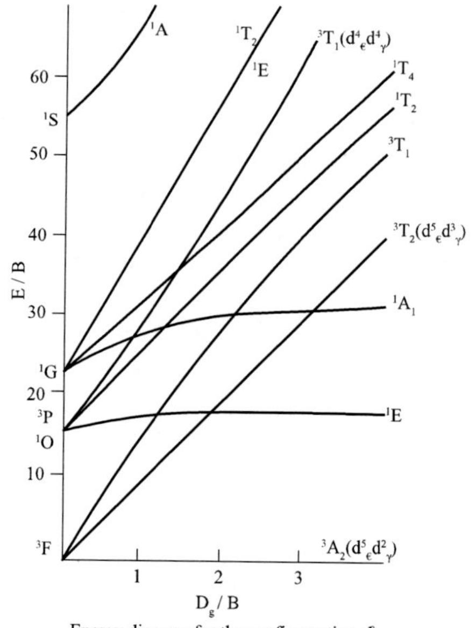

Energy diagram for the configureation d^8

Character Tables for Relevant Symmetry Groups

1. The Nonaxial Groups

C_1	E
A	1

C_2	E	σ_h		
A'	1	1	x, y, R_z	x^2, y^2, z^2, xz
A''	1	−1	z, R_x, R_y	yz, xy

C_i	E	i		
A_9	1	1	R_x, R_y, R_z	x^2, y^2, z^2 xz, xz, yz
A_u	1	−1	x, y, R_z	

2. The C_n Groups

C_2	E	C_2		
A	1	1	$z, \boldsymbol{R}_z$	x^2, y^2, z^2, xy
B	1	−1	$x, y, \boldsymbol{R}_x, \boldsymbol{R}_y$	yz, xz

C_3	E	C_3	C_3^2		$\varepsilon = \exp(2\pi i/3)$
A	1	1	1	$z, \boldsymbol{R}_z$	$x^2 + y^2, z^2$
E	$\left\{\begin{matrix}1 \\ 1\end{matrix}\right.$	$\begin{matrix}\varepsilon \\ \varepsilon^*\end{matrix}$	$\left.\begin{matrix}\varepsilon^* \\ \varepsilon\end{matrix}\right\}$	$(x, y)(\boldsymbol{R}_x, \boldsymbol{R}_y)$	$(x^2 - y^2, xy)(yz, xz)$

C_4	E	C_4	C_2	C_4^3		
A	1	1	1	1	$z, \boldsymbol{R}_z$	$x^2 + y^2, z^2$
B	1	−1	1	−1		$x^2 - y^2, xy$
E	$\left\{\begin{matrix}1 \\ 1\end{matrix}\right.$	$\begin{matrix}i \\ -i\end{matrix}$	$\begin{matrix}-1 \\ -1\end{matrix}$	$\left.\begin{matrix}-i \\ i\end{matrix}\right\}$	$(x, y)(\boldsymbol{R}_x, \boldsymbol{R}_y)$	(yz, xz)

C_5	E	C_5	C_5^2	C_5^3	C_5^4		$\varepsilon = \exp(2\pi i/5)$
A	1	1	1	1	1	$z, \boldsymbol{R}_z$	x^2+y^2, z^2
E_1	$\begin{cases}1\\1\end{cases}$	ε ε^*	ε^2 ε^{2*}	ε^{2*} ε^2	ε^* ε	$(x, y)(\boldsymbol{R}_x, \boldsymbol{R}_y)$	(yz, xz)
E_2	$\begin{cases}1\\1\end{cases}$	ε^2 ε^{2*}	ε^* ε	ε ε^*	ε^{2*} ε^2		(x^2-y^2, xy)

C_6	E	C_6	C_3	C_2	C_3^2	C_6^5		$\varepsilon = \exp(2\pi i/6)$
A	1	1	1	1	1	1	$z, \boldsymbol{R}_z$	x^2+y^2, z^2
B	1	−1	1	−1	1	−1	(x, y)	
E_1	$\begin{cases}1\\1\end{cases}$	ε ε^*	$-\varepsilon^*$ $-\varepsilon$	−1 −1	$-\varepsilon$ $-\varepsilon^*$	ε^* ε	$(\boldsymbol{R}_x, \boldsymbol{R}_y)$	(xz, yz)
E_2	$\begin{cases}1\\1\end{cases}$	$-\varepsilon^*$ $-\varepsilon$	$-\varepsilon$ $-\varepsilon^*$	1 1	$-\varepsilon^*$ $-\varepsilon$	$-\varepsilon$ $-\varepsilon^*$		(x^2-y^2, xy)

C_7	E	C_7	C_7^2	C_7^3	C_7^4	C_7^5	C_7^6		$\varepsilon = \exp(2\pi i/7)$
A	1	1	1	1	1	1	1	$z, \boldsymbol{R}_z$	x^2+y^2, z^2
E_1	$\begin{cases}1\\1\end{cases}$	ε ε^*	ε^2 ε^{2*}	ε^3 ε^{3*}	ε^{3*} ε^3	ε^{2*} ε^2	ε^* ε	$(x, y)(\boldsymbol{R}_x, \boldsymbol{R}_y)$	(xz, yz)
E_2	$\begin{cases}1\\1\end{cases}$	ε^2 ε^{2*}	ε^{3*} ε^3	ε^* ε	ε ε^*	ε^3 ε^{3*}	ε^{2*} ε^2		(x^2-y^2, xy)
E_3	$\begin{cases}1\\1\end{cases}$	ε^3 ε^{3*}	ε^* ε	ε^2 ε^{2*}	ε^{2*} ε^2	ε ε^*	ε^{3*} ε^3		

C_8	E	C_8	C_4	C_2	C_4^3	C_8^3	C_8^5	C_8^7		$\varepsilon = \exp(2\pi i/8)$
A	1	1	1	1	1	1	1	1	$z, \boldsymbol{R}_z$	x^2+y^2, z^2
B	1	−1	1	1	1	−1	−1	−1		
E_1	$\begin{cases}1\\1\end{cases}$	ε ε^*	i $-i$	−1 −1	$-i$ i	$-\varepsilon^*$ $-\varepsilon$	$-\varepsilon$ $-\varepsilon^*$	ε^* ε	(x, y) $(\boldsymbol{R}_x, \boldsymbol{R}_y)$	(xz, yz)
E_2	$\begin{cases}1\\1\end{cases}$	i $-i$	−1 −1	1 1	−1 −1	$-i$ i	i $-i$	$-i$ i		(x^2-y^2, xy)
E_3	$\begin{cases}1\\1\end{cases}$	$-\varepsilon^*$ $-\varepsilon^*$	i $-i$	−1 −1	$-i$ i	ε^* ε	ε ε^*	$-\varepsilon^*$ $-\varepsilon$		

3. The D_n Groups

D_2	E	$C_3(z)$	$C_2(y)$	$C_2(x)$		
A	1	1	1	1		x^2+y^2, z^2
B_1	1	1	−1	−1	z, R_z	xy
B_2	1	−1	1	−1	y, R_y	xz
B_3	1	−1	−1	1	x, R_x	yz

D_3	E	$2C_3$	$3C_2$		
A_1	1	1	1		x^2+y^2, z^2
A_2	1	1	−1	$z, \boldsymbol{R_z}$	
E	2	−1	0	$(x, y)(\boldsymbol{R_x}, \boldsymbol{R_y})$	$(x^2-y^2, xy)(xz, yz)$

D_4	E	$2C_4$	$C2(= C_4^2)$	$2C_2'$	$2C_2''$		
A_1	1	1	1	1	1		x^2+y^2, z^2
A_2	1	1	1	−1	−1	$z, \boldsymbol{R_z}$	
B_1	1	−1	1	1	−1		x^2-y^2
B_2	1	−1	1	−1	1		xy
E	2	0	−2	0	0	$(x, y)(\boldsymbol{R_x}, \boldsymbol{R_y})$	(xz, yz)

D_5	E	$2C_5$	$2C_5^2$	$5C_2$		
A_1	1	1	1	1		x^2+y^2, z^2
A_2	1	1	1	−1	$z, \boldsymbol{R_z}$	
E_1	2	2 cos 72°	2 cos 144°	0	$(x, y)(\boldsymbol{R_x}, \boldsymbol{R_y})$	(xz, yz)
E_2	2	2 cos 144°	2 cos 72°	0		$(x^2-y^2,\ xy)$

D_6	E	$2C_6$	$2C_3$	C_2	$3C_2'$	$3C_2''$		
A_1	1	1	1	1	1	1		x^2+y^2, z^2
A_2	1	1	1	1	−1	−1	$z, \boldsymbol{R_z}$	
B_1	1	−1	1	−1	1	−1		
B_2	1	−1	1	−1	−1	1		
E_1	2	1	−1	−2	0	0	$(x, y)(\boldsymbol{R_x}, \boldsymbol{R_y})$	(xz, yz)
E_2	2	−1	−1	2	0	0		$(x^2-y^2,\ xy)$

4. The C_{nv} Groups

C_{2v}^{a}	E	C_2	$\sigma_v(xz)$	$\sigma_v'(yz)$		
A_1	1	1	1	1	z	x^2, y^2, z^2
A_2	1	1	−1	−1	$\boldsymbol{R_z}$	xy
B_1	1	−1	1	−1	x, $\boldsymbol{R_y}$	xz
B_2	1	−1	−1	1	y, $\boldsymbol{R_x}$	yz

[a]For a planar molecule the x-axis is taken perpendicular to the plane

C_{3v}	E	$2C_3$	$3\sigma_v$		
A_1	1	1	1	z	$x^2 + y^2$, z^2
A_2	1	1		$\boldsymbol{R_z}$	
E	2	−1	0	$(x, y)(\boldsymbol{R_x}, \boldsymbol{R_y})$	$(x^2 - y^2, xy)(xz, yz)$

C_{4v}^{a}	E	$2C_4$	C_2	$2\sigma_v$	$2\sigma_d$		
A_1	1	1	1	1	1	z	$x^2 + y^2$, z^2
A_2	1	1	1	−1	−1	$\boldsymbol{R_z}$	
B_1	1	−1	1	1	−1		$x^2 - y^2$
B_2	1	−1	1	−1	1		xy
E	2	0	−2	0	0	$(x, y)(\boldsymbol{R_x}, \boldsymbol{R_y})$	(xz, yz)

C_{5v}	E	$2C_5$	$2C_5^2$	$5\sigma_v$		
A_1	1	1	1	1	z	$x^2 + y^2, z^2$
A_2	1	1	1	−1	$\boldsymbol{R_z}$	
E_1	2	2 cos 72°	2 cos 144°	0	$(x, y)(\boldsymbol{R_x}, \boldsymbol{R_y})$	(xz, yz)
E_2	2	2 cos 144°	2 cos 72°	0		$(x^2 - y^2, xy)$

C_{6v}	E	$2C_6$	$2C_3$	C_2	$3\sigma_v$	$3\sigma_d$		
A_1	1	1	1	1	1	1	z	$x^2 + y^2, z^2$
A_2	1	1	1	1	−1	−1	$\boldsymbol{R_z}$	
B_1	1	−1		−1		−1		
B_2	1	−1		−1	−1	1		
E_1	2	1	−1	−2	0	0	$(x, y)(\boldsymbol{R_x}, \boldsymbol{R_y})$	(xz, yz)
E_2	2	−1	−1	2	0	0		$(x^2 - y^2, xy)$

5. The C_{nh} Groups

C_{2h}	E	C_2	i	σ_h		
A_g	1	1	1	1	$\boldsymbol{R_z}$	x^2, y^2, z^2, xy
B_g	1	−1	1	−1	$\boldsymbol{R_x}, \boldsymbol{R_y}$	xz, yz
A_u	1	1	−1	−1	z	
B_u	1	−1	−1	1	x, y	

C_{3h}	E	C_3	C_3^2	σ_h	S_3	S_3^5		$\varepsilon = \exp(2\pi i/3)$
A'	1	1	1	1	1	1	$\boldsymbol{R_z}$	x^2+y^2, z^2
E'	1	ε	ε^*	1	ε	ε^*	(x, y)	(x^2-y^2, xy)
	1	ε^*	ε	1	ε^*	ε		
A''	1	1	1	−1	−1	−1	z	
E''	1	ε	ε^*	−1	$-\varepsilon$	$-\varepsilon^*$	$(\boldsymbol{R_x}, \boldsymbol{R_y})$	(xz, yz)
	1	ε^*	ε	−1	$-\varepsilon^*$	$-\varepsilon$		

C_{4h}	E	C_4	C_2	C_4^3	i	S_4^3	σ_h	S_4		
A_g	1	1	1	1	1	1	1	1	$\boldsymbol{R_z}$	x^2+y^2, z^2
B_g	1	−1	1	−1	1	−1	1	−1		x^2-y^2, xy
E_g	1	i	−1	$-i$	1	i	−1	$-i$	$(\boldsymbol{R_x}, \boldsymbol{R_y})$	(xz, yz)
	1	$-i$	−1	i	1	$-i$	−1	i		
A_u	1	1	1	1	−1	−1	−1	−1	z	
B_u	1	−1	1	−1	−1	1	−1	1		
E_u	1	i	−1	$-i$	−1	$-i$	1	i	(x, y)	
	1	$-i$	−1	i	−1	i	1	$-i$		

C_{5h}	E	C_3	C_5^2	C_5^3	C_5^4	σ_h	S_5	S_5^7	S_5^3	S_5^9		$\varepsilon = \exp(2\pi i/5)$
A'	1	1	1	1	1	1	1	1	1	1	$\boldsymbol{R_z}$	x^2+y^2, z^2
E_1'	1	ε	ε^2	ε^{2*}	ε^*	1	ε	ε^2	ε^{2*}	ε^*	(x, y)	
	1	ε^*	ε^{2*}	ε^2	ε	1	ε^*	ε^{2*}	ε^2	ε		
E_2'	1	ε^2	ε^*	ε	ε^{2*}	1	ε^2	ε^*	ε	ε^{2*}		(x^2-y^2, xy)
	1	ε^{2*}	ε	ε^*	ε^2	1	ε^{2*}	ε	ε^*	ε^2		
A''	1	1	1	1	1	−1	−1	−1	−1	−1	z	
E_1''	1	ε	ε^2	ε^{2*}	ε^*	−1	$-\varepsilon$	$-\varepsilon^2$	$-\varepsilon^{*2}$	$-\varepsilon^*$	$(\boldsymbol{R_x}, \boldsymbol{R_y})$	(xz, yz)
	1	ε^*	ε^{2*}	ε^2	ε	−1	$-\varepsilon^*$	$-\varepsilon^{2*}$	$-\varepsilon^2$	$-\varepsilon$		
E_2''	1	ε^2	ε^*	ε	ε^{2*}	−1	$-\varepsilon^2$	$-\varepsilon^*$	$-\varepsilon$	$-\varepsilon^{2*}$		
	1	ε^{2*}	ε	ε^*	ε^2	−1	$-\varepsilon^{2*}$	$-\varepsilon$	$-\varepsilon^*$	$-\varepsilon^2$		

C_{6h}	E	C_6	C_3	C_2	C_3^2	C_6^5	i	S_3^5	S_6^5	σ_h	S_5	S_3		$\varepsilon = \exp(2\pi i/6)$
A_g	1	1	1	1	1	1	1	1	1	1	1	1	$\boldsymbol{R_z}$	x^2+y^2, z^2
B_g	1	−1	1	−1	1	−1	1	−1	1	−1	1	−1		
E_{1g}	1 1	ε ε^*	ε^2 ε^{2*}	−1 −1	$-\varepsilon$ $-\varepsilon^*$	ε^* ε	1 1	ε ε^*	$-\varepsilon^*$ $-\varepsilon$	−1 −1	$-\varepsilon$ $-\varepsilon^*$	ε^* ε	$(\boldsymbol{R_x}, \boldsymbol{R_y})$	(xz, yz)
E_{2g}	1 1	$-\varepsilon^*$ $-\varepsilon$	ε^* ε	1 1	$-\varepsilon^*$ $-\varepsilon$	$-\varepsilon$ $-\varepsilon^*$	1 1	$-\varepsilon^*$ $-\varepsilon$	$-\varepsilon$ $-\varepsilon^*$	1 1	$-\varepsilon^*$ $-\varepsilon$	$-\varepsilon$ $-\varepsilon^*$		(x^2-y^2, xy)
A_u	1	1	1	1	1	1	−1	−1	−1	−1	−1	−1	z	
B_u	1	−1	1	−1	1	−1	−1	1	−1	1	−1	1		
E_{1u}	1 1	ε ε^*	ε^2 ε^{2*}	−1 −1	$-\varepsilon$ $-\varepsilon^*$	ε^* ε^*	−1 −1	$-\varepsilon$ $-\varepsilon^*$	ε^* ε	1 1	ε ε^*	$-\varepsilon^*$ $-\varepsilon$	(x, y)	
E_{2u}	1 1	$-\varepsilon^*$ $-\varepsilon$	ε^* ε	1 1	$-\varepsilon^*$ $-\varepsilon$	$-\varepsilon$ $-\varepsilon^*$	−1−1	ε^* ε	ε ε^*	−1 −1	ε^* ε	ε ε^*		

6. The D_{nh} Groups

D_{2h}	E	$C_2(z)$	$C_2(y)$	$C_2(x)$	i	$\sigma(xy)$	$\sigma(xz)$	$\sigma(yz)$		
A_g	1	1	1	1	1	1	1	1		x^2, y^2, z^2
B_{1g}	1	1	−1	−1	1	1	−1	−1	$\boldsymbol{R_z}$	xy
B_{2g}	1	−1	1	−1	1	−1	1	−1	$\boldsymbol{R_y}$	xz
B_{3g}	1	−1	−1	1	1	−1	−1	1	$\boldsymbol{R_x}$	yz
A_u	1	1	1	1	−1	−1	−1	−1		
B_{1u}	1	1	−1	−1	−1	−1	1	1	z	
B_{2u}	1	−1	1	−1	−1	1	−1	1	y	
B_{3u}	1	−1	−1	1	−1	1	1	−1	x	

D_{3h}	E	$2C_3$	$3C_2$	σ_h	$2S_3$	$3\sigma_v$		
A_1'	1	1	1	1	1	1		x^2+y^2, z^2
A_2'	1	1	−1	1	1	−1	$\boldsymbol{R_z}$	
E'	2	−1	0	2	−1	0	(x, y)	(x^2-y^2, xy)
A_1''	1	1	1	−1	−1	−1		
A_2''	1	1	−1	−1	−1	1	z	
E''	2	−1	0	−2	1	0	$(\boldsymbol{R_x}, \boldsymbol{R_y})$	(xz, yz)

D_{4h}^a	E	$2C_4$	C_2	$2C_2'$	$2C_2''$	i	$2S_4$	σ_h	$2\sigma_v$	$2\sigma_d$		
A_{1g}	1	1	1	1	1	1	1	1	1	1		x^2+y^2, z^2
A_{2g}	1	1	1	−1	−1	1	1	1	−1	−1	$\boldsymbol{R_z}$	
B_{1g}	1	−1	1	1	−1	1	−1	1	1	−1		x^2-y^2
B_{2g}	1	−1	1	−1	1	1	−1	1	−1	1		xy
E_g	2	0	−2	0	0	2	0	−2	0	0	$(\boldsymbol{R_x}, \boldsymbol{R_y})$	(xz, yz)
A_{1u}	1	1	1	1	1	−1	−1	−1	−1	−1		
A_{2u}	1	1	1	−1	−1	−1	−1	−1	1	1	z	

(continued)

(continued)

D_{4h}^{a}	E	$2C_4$	C_2	$2C_2'$	$2C_2''$	i	$2S_4$	σ_h	$2\sigma_v$	$2\sigma_d$		
B_{1u}	1	−1	1	1	−1	−1	1	−1	−1	1		
B_{2u}	1	−1	1	−1	1	−1	1	−1	1	−1		
E_u	2	0	−2	0	0	−2	0	2	0	0	(x, y)	

[a] σ_a passes through the atoms and σ_d bisects the bond angles

D_{5h}	E	$2C_5$	$2C_5^2$	$5C_2$	σ_h	$2S_5$	$2S_5^3$	$5\sigma_h$		
A_1'	1	1	1	1	1	1	1	1		x^2+y^2, z^2
A_2'	1	1	1	−1	1	1	1	−1	R_z	
E_1'	2	2 cos 72°	2 cos 144°	0	2	2 cos 72°	2 cos 144°	0	(x, y)	
E_2'	2	2 cos 144°	2 cos 72°	0	2	2 cos 144°	2 cos 72°	0		(x^2-y^2, xy)
A_1''	1	1	1	1	−1	−1	−1	−1		
A_2''	1	1	1	−1	−1	−1	−1	1	z	
E_1''	2	2 cos 72°	2 cos 144°	0	−2	−2 cos 72°	−2 cos 144°	0	(R_x, R_y)	(xz, yz)
E_2''	2	2 cos 144°	2 cos 72°	0	−2	−2 cos 144°	−2 cos 72°	0		

D_{6h}	E	$2C_6$	$2C_3$	C_2	$3C_2'$	$3C_2''$	i	$2S_3$	$2S_6$	σ_h	$3\sigma_d$	$3\sigma_v$		
A_{1g}	1	1	1	1	1	1	1	1	1	1	1	1		x^2+y^2, z^2
A_{2g}	1	1	1	1	−1	−1	1	1	1	1	−1	−1	R_z	
B_{1g}	1	−1	1	−1	1	−1	1	−1	1	−1	1	−1		
B_{2g}	1	−1	1	−1	−1	1	1	−1	1	−1	−1	1		
E_{1g}	2	1	−1	−2	0	0	2	1	−1	−2	0	0	(R_x, R_y)	(xz, yz)
E_{2g}	2	−1	−1	2	0	0	2	−1	−1	2	0	0		(x^2-y^2, xy)
A_{1u}	1	1	1	1	1	1	−1	−1	−1	−1	−1	−1		
A_{2u}	1	1	1	1	−1	−1	−1	−1	−1	−1	1	1	z	
B_{1u}	1	−1	1	−1	1	−1	−1	1	−1	1	−1	1		
B_{2u}	1	−1	1	−1	−1	1	−1	1	−1	1	1	−1		
E_{1u}	2	1	−1	−2	0	0	−2	−1	1	2	0	0	(x, y)	
E_{2u}	2	−1	−1	2	0	0	−2	1	1	−2	0	0		

D_{8h}	E	$2C_8$	$2C_8^3$	$2C_4$	C_2	$4C_2'$	$4C_2''$	i	$2S_8$	$2S_8^3$	$2S_4$	σ_h	$4\sigma_d$	$4\sigma_v$		
A_{1g}	1	1	1	1	1	1	1	1	1	1	1	1	1	1		x^2+y^2, z^2
A_{2g}	1	1	1	1	1	−1	−1	1	1	1	1	1	−1	−1	R_z	
B_{1g}	1	−1	−1	1	1	1	−1	1	−1	−1	1	1	1	−1		
B_{2g}	1	−1	−1	1	1	−1	1	1	−1	−1	1	1	−1	1		
E_{1g}	2	$\sqrt{2}$	$-\sqrt{2}$	0	−2	0	0	2	$\sqrt{2}$	$-\sqrt{2}$	0	−2	0	0	R_x, R_y	(xz, yz)
E_{2g}	2	0	0	−2	2	0	0	2	0	0	−2	2	0	0		(x^2-y^2, xy)
E_{3g}	2	$-\sqrt{2}$	$\sqrt{2}$	0	−2	0	0	2	$-\sqrt{2}$	$\sqrt{2}$	0	−2	0	0		
A_{1u}	1	1	1	1	1	1	1	−1	−1	−1	−1	−1	−1	−1		
A_{2u}	1	1	1	1	−1	−1	−1	−1	−1	−1	−1	−1	1	1	z	
B_{1u}	1	−1	−1	1	1	1	−1	−1	1	1	−1	−1	−1	1		
B_{2u}	1	−1	−1	1	1	−1	1	−1	1	1	−1	−1	1	−1		
E_{1u}	2	$\sqrt{2}$	$-\sqrt{2}$	0	−2	0	0	−2	$-\sqrt{2}$	$\sqrt{2}$	0	2	0	0	(x, y)	
E_{2u}	2	0	0	−2	2	0	0	−2	0	0	2	−2	0	0		
E_{3u}	2	$-\sqrt{2}$	$\sqrt{2}$	0	−2	0	0	−2	$\sqrt{2}$	$-\sqrt{2}$	0	2	0	0		

7. The D_{nd} Groups

D_{2d}	E	$2S_4$	C_2	$2C_2'$	$2\sigma_d$		
A_1	1	1	1	1	1		x^2+y^2, z^2
A_2	1	1	1	−1	−1	$\boldsymbol{R}_z$	
B_1	1	−1	1	1	−1		x^2-y^2
B_2	1	−1	1	−1	1	z	xy
E	1	0	−2	0	0	(x, y), $(\boldsymbol{R}_x, \boldsymbol{R}_y)$	(xz, yz)

D_{3d}	E	$2C_3$	$3C_2$	i	$2S_6$	$3\sigma_d$		
A_{1g}	1	1	1	1	1	1		x^2+y^2, z^2
A_{2g}	1	1	−1	1	1	−1	$\boldsymbol{R}_z$	
E_g	2	−1	0	2	−1	0	$(\boldsymbol{R}_x, \boldsymbol{R}_y)$	$(x^2-y^2,\ xy)$ (xz, yz)
A_{1u}	1	1	1	−1	−1	−1		
A_{2u}	1	1	−1	−1	−1	1	z	
E_u	2	−1	0	−2	1	0	(x, y)	

D_{4d}	E	$2S_8$	$2C_4$	$2S_8^3$	C_2	$4C_2'$	$4\sigma_d$		
A_1	1	1	1	1	1	1	1		x^2+y^2, z^2
A_2	1	1	1	1	1	−1	−1	$\boldsymbol{R}_z$	
B_1	1	−1	1	−1	1	1	−1		
B_2	1	−1	1	−1	1	−1	1	z	
E_1	2	$\sqrt{2}$	0	$-\sqrt{2}$	−2	0	0	(x, y)	
E_2	2	0	−2	0	2	0	0		$(x^2-y^2,\ xy)$
E_3	2	$-\sqrt{2}$	0	$\sqrt{2}$	−2	0	0	$(\boldsymbol{R}_x, \boldsymbol{R}_y)$	(xz, yz)

D_{5d}	E	$2C_5$	$2C_5^2$	$5C_2$	i	$2S_{10}^3$	$2S_{10}$	$5\sigma_h$		
A_{1g}	1	1	1	1	1	1	1	1		x^2+y^2, z^2
A_{2g}	1	1	1	−1	1	1	1	−1	$\boldsymbol{R}_z$	
E_{1g}	2	2 cos 72°	2 cos 144°	0	2	2 cos 72°	2 cos 144°	0	$(\boldsymbol{R}_x, \boldsymbol{R}_y)$	$(xz,\ yz)$
E_{2g}	2	2 cos 144°	2 cos 72°	0	2	2 cos 144°	2 cos 72°	0		$(x^2-y^2,\ xy)$
A_{1u}	1	1	1	1	−1	−1	−1	−1		
A_{2u}	1	1	1	−1	−1	−1	−1	1	z	
E_{1u}	2	2 cos 72°	2 cos 144°	0	−2	−2 cos 72°	−2 cos 144°	0	$(x,\ y)$	
E_{2u}	2	2 cos 144°	2 cos 72°	0	−2	−2 cos 144°	−2 cos 72°	0		

D_{6d}	E	$2S_{12}$	$2C_4$	$2S_4$	$2C_3$	$2S_{12}^5$	C_2	$6C_2'$	$6\sigma_d$		
A_1	1	1	1	1	1	1	1	1	1		x^2+y^2, z^2
A_2	1	1	1	1	1	1	1	–1	–1	$\boldsymbol{R}_z$	
B_1	1	–1	1	–1	1	–1	1	1	–1		
B_2	1	–1	1	–1	1	–1	1	–1	1	z	
E_1	2	$\sqrt{3}$	1	0	–1	$-\sqrt{3}$	–2	0	0	(x, y)	
E_2	2	1	–1	–2	–1	1	2	0	0		(x^2-y^2, xy)
E_3	2	0	–2	0	2	0	–2	0	0		
E_4	2	–1	–1	2	–1	–1	2	0	0		
E_5	2	$-\sqrt{3}$	1	0	–1	$\sqrt{3}$	–2	0	0	$(\boldsymbol{R}_x, \boldsymbol{R}_y)$	(xz, yz)

8. The S_n Groups

S_4	E	S_4	C_2	S_4^3		
A	1	1	1	1	$\boldsymbol{R}_z$	x^2+y^2, z^2
B	1	–1	1	–1	z	x^2-y^2, xy
E	1	i	–1	$-i$	$(x, y); (\boldsymbol{R}_x, \boldsymbol{R}_y)$	(xz_z, yz_z)
	1	$-i$	–1	i		

S_6	E	C_3	C_3^2	i	S_6^5	S_6		$\varepsilon = \exp(2\pi i/3)$
A_g	1	1	1	1	1	1	$\boldsymbol{R}_z$	x^2+y^2, z^2
E_s	1	ε	ε^*	1	ε	ε^*	$(\boldsymbol{R}_x, \boldsymbol{R}_y)$	(x^2-y^2, xy);
	1	ε^*	ε	1	ε^*	ε		
A_g	1	1	1	–1	–1	–1	z	(xz, yz)
E_s	1	ε	ε^*	–1	$-\varepsilon$	$-\varepsilon^*$	(x, y)	
	1	ε^*	ε	–1	$-\varepsilon^*$	$-\varepsilon$		

S_8	E	S_8	C_4	S_5^3	C_2	S_8^5	C_4^3	S_8^7		$\varepsilon = \exp(2\pi i/8)$
A	1	1	1	1	1	1	1	1	$\boldsymbol{R}_z$	$x^2+y^2, z^2x^2+y^2, z^2$
B	1	–1	1	–1	1	–1	1	–1	z	
E_1	1	ε	i	$-\varepsilon^*$	–1	$-\varepsilon$	$-i$	ε^*	$(x, y), (\boldsymbol{R}_x, \boldsymbol{R}_y)$	
	1	ε^*	$-i$	$-\varepsilon$	–1	$-\varepsilon^*$	i	ε		
E_2	1	i	–1	$-i$	1	i	–1	$-i$		(x^2-y^2, xy)
	1	$-i$	–1	i	1	$-i$	–1	i		
E_3	1	$-\varepsilon^*$	$-i$	ε	–1	ε^*	i	$-\varepsilon$		(xz, yz)
	1	$-\varepsilon$	i	ε^*	–1	ε	$-i$	$-\varepsilon^*$		

9. The Cubic Groups

T	E	$4C_3$	$4C_5^2$	$3C_2$		$\varepsilon = \exp(2\pi i/3)$
A	1	1	1	1		$x^2+y^2+z^2$
E	$\begin{cases}1\\1\end{cases}$	ε ε^*	ε^* ε	1 1		$(2z^2-x^2-y^2, x^2-y^2)$
T	3	0	0	-1	$(\boldsymbol{R_x}, \boldsymbol{R_y}, \boldsymbol{R_z})$; (x, y, z)	(xy, xz, yz)

T_h	E	$4C_3$	$4C_5^2$	$3C_3$	i	$4S_6$	$4S_5^7$	$3\sigma_k$		$\varepsilon = \exp(2\pi i/3)$
A_g	1	1	1	1	1	1	1	1		$x^2+y^2+z^2$
A_sS	1	1	1	1	-1	-1	-1	-1		
E_g	$\begin{cases}1\\1\end{cases}$	ε ε^*	ε^* ε	1 1	1 1	ε ε^*	ε^* ε	1 1		$(2z^2-x^2-y^2, x^2-y^2)$
E_u	$\begin{cases}1\\1\end{cases}$	ε ε^*	ε^* ε	1 1	-1 -1	$-\varepsilon$ $-\varepsilon^*$	$-\varepsilon^*$ $-\varepsilon$	-1 -1		
T_g	3	0	0	-1	1	0	0	-1	$(\boldsymbol{R_x}, \boldsymbol{R_y}, \boldsymbol{R_z})$	(xz, yz, xy)
T_u	3	0	0	-1	-1	0	0	1	(x, y, z)	

T_d	E	$8C_3$	$3C_2$	$6S_4$	$6\sigma_d$		
A_1	1	1	1	1	1		$x^2+y^2+z^2$
A_2	1	1	1	-1	-1		
E	2	-1	2	0	0		$(2z^2-x^2-y^2,\ x^2-y^2)$
T_1	3	0	-1	1	-1	$(\boldsymbol{R_x}, \boldsymbol{R_y}, \boldsymbol{R_z})$	
T_2	3	0	-1	-1	1	$(x,\ y,\ z)$	(xy, xz, yz)

O	E	$6C_4$	$3C_2(=C_4^2)$	$8C_3$	$6C_3$		
A_1	1	1	1	1	1		$x^2+y^2+z^2$
A_2	1	-1	1	1	-1		
E	2	0	2	-1	0		$(2z^2-x^2-y^2,\ x^2-y^2)$
T_1	3	1	-1	0	-1	$(\boldsymbol{R_x}, \boldsymbol{R_y}, \boldsymbol{R_z}, x, y, z)$	
T_2	3	-1	-1	0	1		(xy, xz, yz)

O_4	E	$8C_3$	$6C_2$	$6C_4$	$3C_2(=C_4^2)$	i	$6S_4$	$6S_6$	$3\sigma_h$	$6\sigma_d$		
A_{1g}	1	1	1	1	1	1	1	1	1	1		$x^2+y^2+z^2$
A_{2g}	1	1	−1	−1	1	1	−1	1	1	−1		
E_g	2	−1	0	0	2	2	0	−1	2	0		$(2z^2-x^2-y^2,\ x^2-y^2)$
T_{1g}	3	0	−1	1	−1	3	1	0	−1	−1	$(\boldsymbol{R_x}, \boldsymbol{R_y}, \boldsymbol{R_z})$	
T_{2g}	3	0	1	−1	−1	3	−1	0	−1	1		(xy, xz, yz)
A_{1u}	1	1	1	1	1	−1	−1	−1	−1	−1		
A_{2u}	1	1	−1	−1	1	−1	1	−1	−1	1		
E_u	2	−1	0	0	2	−2	0	1	−2	0		
T_{1u}	3	0	−1	1	−1	−3	−1	0	1	1	(x, y, z)	
T_{2u}	3	0	1	−1	−1	−3	1	0	1	−1		

10. The Groups $C_{\infty v}$, and $D_{\infty h}$ for Linear Molecules

$C_{\infty v}$	E	$2C_\infty^\Phi$	...	$\infty\sigma_v$		
$A_1 \equiv \Sigma^+$	1	1	...	1	z	x^2+y^2, z^2
$A_2 \equiv \Sigma^-$	1	1	...	−1	$\boldsymbol{R_z}$	
$E_1 \equiv \Pi$	2	$2\cos\Phi$	...	0	$(x, y); (\boldsymbol{R_x}, \boldsymbol{R_y})$	(xz, yz)
$E_2 = \Delta$	2	$2\cos 2\Phi$	...	0		(x^2-y^2, xy)
$E_3 \equiv \Phi$	2	$2\cos 3\Phi$	...	0		
...	...	...	...	...		

$D_{x,h}$	E	$2C_\infty^\Phi$	...	$\infty\sigma_v$	i	$2S_\infty^\Phi$	...	∞C_2		
Σ_g^+	1	1	...	1	1	1	...	1		x^2+y^2, z^2
Σ_g^-	1	1	...	−1	1	1	...	−1	$\boldsymbol{R_z}$	
Π_g'	2	$2\cos\Phi$	...	0	2	$-2\cos\Phi$	...	0	$(\boldsymbol{R_x}, \boldsymbol{R_y})$	(xz, yz)
Δ_g	2	$2\cos 2\Phi$	...	0	2	$2\cos 2\Phi$	...	0		(x^2-y^2, xy)
...	...	...	...	...	...	...	...	...		
Σ_∞^+	1	1	...	1	−1	−1	...	−1	z	
Σ_∞^-	1	1	...	−1	−1	−1	...	1		
Π_∞'	2	$2\cos\Phi$	...	0	−2	$2\cos\Phi$	...	0	(x, y)	
Δ_∞	2	$2\cos 2\Phi$	...	0	−2	$-2\cos 2\Phi$	...	0		
...	...	...	...	...	...	...	...	...		

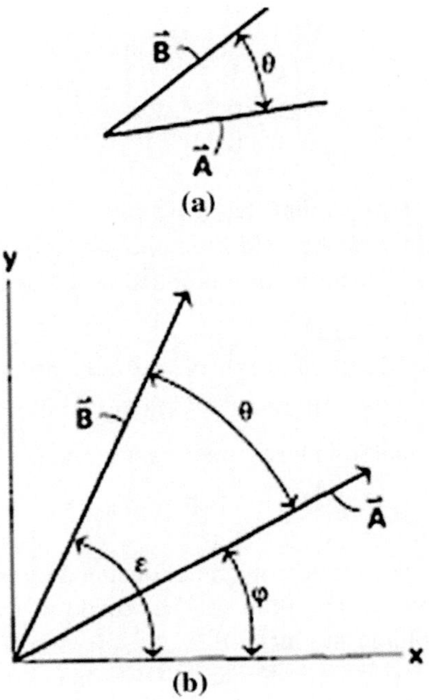

Fig. A.1 Vector **A** and **B** separated by an angle **θ** and placed in an **xy** space

In a p dimension space

$$\mathbf{A} \cdot \mathbf{B} = \sum_{i=1}^{p} A_i B_i$$

Properties of Matrices

A matrix is an array of numbers or symbols with the following general form

$$\begin{bmatrix} a_{11} & a_{12} & a_{13} \\ a_{21} & a_{22} & a_{23} \\ a_{31} & a_{32} & a_{33} \end{bmatrix}$$

The square brackets indicate the matrix, as opposite to a determinant. It can be used $[a_{ij}]$ as a symbol, meaning that the matrix has i rows and j columns. a_{ij} is an element of the ith row and jth column. If the number of rows and columns is equal, the matrix is called square. In a square matrix, the elements a_{ij}, where $i = j$, are called diagonal elements, the others are called off-diagonal elements. If these later are 0, the matrix is called diagonal. If the diagonal elements are 1, the matrix is called unit matrix

$$\begin{bmatrix} 1 & 0 & 0 & 0 \\ 0 & 1 & 0 & 0 \\ 0 & 0 & 1 & 0 \\ 0 & 0 & 0 & 1 \end{bmatrix} = \delta$$

The trace or character of a diagonal matrix is the sum of the diagonal elements. A vector is conveniently represented by a one row or column matrix.

Matrices can be added subtracted multiplied or divided by using the rules of the matrix algebra.

(1) We can add or subtract matrices with the same number of rows and columns. For matrices $[a_{ij}]$ and $[b_{ij}]$, the matrix sum $[c_{ij}]$ has elements $c_{ij} = a_{ij} + b_{ij}$

(2) A matrix can be multiplied by a scalar, multiplying each element by the scalar

$$k[a_{ij}] = [ka_{ij}]$$

(3) The c_{ij} element of a product matrix is obtained by multiplying the *ith* row of the first matrix by the *jth* column of the second matrix, and it is consequently called row by column product.

$c_{ik} = {}^{n}\sum_{j=1} a_{ij}\, b_{jk}$. In order to multiply two matrices, the number of columns of $[a_{ij}]$ must be equal to that of rows of $[b_{ij}]$.

$$\underset{3 \text{ by } 2}{\begin{bmatrix} a_{11} & a_{12} \\ a_{21} & a_{22} \\ a_{31} & a_{32} \end{bmatrix}} \underset{2 \text{ by } 3}{\begin{bmatrix} b_{11} & b_{12} & b_{13} \\ b_{21} & b_{22} & b_{23} \end{bmatrix}} = \underset{3 \text{ by } 3}{\begin{bmatrix} c_{11} & c_{12} & c_{13} \\ c_{21} & c_{22} & c_{23} \\ c_{31} & c_{32} & c_{33} \end{bmatrix}}$$

$$c_{11} = a_{11}b_{11} + a_{12}b_{21}$$
$$c_{12} = a_{11}b_{12} + a_{12}b_{22}$$
$$c_{13} = a_{11}b_{13} + a_{12}b_{23}$$
$$c_{21} = a_{21}b_{11} + a_{22}b_{21}$$
$$c_{22} = a_{21}b_{12} + a_{22}b_{22}$$
$$c_{23} = a_{21}b_{13} + a_{22}b_{23}$$
$$c_{31} = a_{31}b_{11} + a_{32}b_{21}$$
$$c_{32} = a_{31}b_{12} + a_{32}b_{22}$$
$$c_{33} = a_{31}b_{13} + a_{32}b_{23}$$

(4) The matrix multiplication obeys to the associative law

(5) The division of matrices is based on the fact that $[a_{ij}]/[b_{ij}] = [a_{ij}].[b_{ij}]^{-1}$, where $[b_{ij}]^{-1}$ is the inverse matrix and $[b_{ij}] \cdot [b_{ij}]^{-1} = 1$ (unit matrix)

(6) The symmetry operation can be represented by matrices

(7) Two matrices are called conjugated when they are related by a similarity transformation through a third matrix $[r_{ij}]$ in the way as

$$[a_{jj}] = [r_{ij}]^{-1}[b_{jj}][r_{ij}].$$

References

1. R.S. Drago, Physical Methods in Chemistry (Ed Saunders College Publishing, 1977), pp. 70, 382
2. J.A. Weil, J.R. Bolton, J.E. Wertz, Electron Paramagnetic Resonance: Elementary Theory and Practical Application, 1st ed. (Wiley, 1994), pp. 8, 11, 13
3. G. Longoni, F. Morazzoni, Electron spin resonance behaviour of the $[Fe_3Pt_3\ (CO)_{15}]^-$ Carbonyl cluster anion. J.C.S. Dalton 1735–1737 (1981)
4. T. Beringhelli, F. Morazzoni, D. Strumolo, ESR investigation of paramagnetic carbonyl—metal clusters of high nuclearity. J. Organomet. Chem. **236**, 109–118 (1982)
5. F.Cariati, F. Morazzoni,C. Busetto, G. Del Piero, A. Zazzetta, Paramagneti anisotropy in cobalt(II) schiff—base complexes. X-Ray crystal structure and electron spin resonance of NN′ Ethylenebis-(acetylacetoneiminato)cobalt(II) doped NN′ Ethylenebis-(acetylacetoneiminato)nickel(II). J.C.S. Dalton 342–347 (1981)
6. W.P. Schaefer, R.E. Marsh, Oxygen-carrying cobalt compounds. I. Bis(salicylaldehyde) ethylenediiminecobalt(II) monochloroformate. Acta Cryst. B25, 1675 (1969)
7. M. Calligaris, D. Minichelli, G. Nardin, L. Randaccio, Structural aspects of the synthetic oxygen-carrier NN′ Ethylenebis-(acetylacetoneiminato)cobalt(II). Part II. crystal and molecular structure of the monopyridine complex. J.C.S. A 2411–2415 (1970)
8. C. Busetto, F. Cariati, A. Fusi, M. Gullotti, F. Morazzoni, A. Pasini, R. Ugo, V. Valenti, Optically active complexes of Schiff bass. Part II. Complexes of cobalt(II) with tetradentate Schiff bases and their reactivity with oxygen. J.C.S. Dalton 754–765 (1973)
9. F. Cariati, D. Galizzioli, F. Morazzoni, New adducts of phthalocyaninatocobalt(II) with pyridine and 4-methylpyridine and their vibrational, magnetic and electronic properties. Part I. Reactivity towards oxygen. J.C.S. Dalton 556–561 (1975)
10. F. Cariati, G. Micera, M.A. Zoroddu, F. Morazzoni, New adducts of phthalocyaninatocobalt (II) with base molecules. Part 3. Chemical and spectroscopic relations between mono (Low- and high-spin) and Bis adducts. J.C.S. Dalton 535–538 (1981)
11. F. Campadelli, F. Cariat, P. Carniti, F. Morazzoni, V. Ragaini, Oxygen reduction on γAl_2O_3 surfaces by electron transfer from porphyrin-like complexes. J. Catalysis **44**,167–169 (1976)
12. T. Beringhelli, A. Gervasini, F. Morazzoni, D. Strumolo, Paramagnetic metal and oxygen species observed with $[Co_4(CO)_{12}]$ and $[Rh_4(CO)_{12}]$ carbonyl clusters pyrolysed on γ. alumina and zirconia supports. J. Chem. Soc. Faraday Trans. I **80**, 1479–1489 (1984)
13. S. Pizzini, D. Narducci, D. Daverio, C.M. Mari, F. Morazzoni, A. Gervasini, Optical and spectromagnetical properties of phosphate glasses containing ruthenium and titanium ions. J. Chem. Soc. Faraday Trans. I **83**, 705–719 (1987)

F. Morazzoni, *Symmetry in Inorganic and Coordination Compounds*, Lecture Notes in Chemistry 106, https://doi.org/10.1007/978-3-030-62996-0

14. V. Malatesta, F. Morazzoni, L. Pellicciari-Bollini, R. Scotti, The interaction between superoxide dismutase and doxorubicin: An electron spin resonance approach. J. Chem. Soc. Faraday Trans. I **83**, 3669–3673 (1987)
15. E. Monti, F. Morazzoni, G. Perletti, F. Piccinini, A new approach to the direct dtection of free radical in the intact myocardium. Free Rad. Res. Comms. **8**, 161–166 (1989)
16. F. Morazzoni, R. Scotti, S. Volonté, Electron Paramagnetic Resonance Investigation of Paramagnetic Point Defects in ZnO and ZnO-supported Ruthenium. J. Chem. Soc. Faraday Trans. **86**, 1587–1591 (1990)
17. P. Di Nola, F. Morazzoni, R. Scotti, D. Narducci, Paramagnetic point defects in SnO_2 and their reactivity with the surrounding gases. J. Chem. Soc. Faraday Trans. **89**, 3711–3713 (1993)
18. I. Jiménez, M.A. Centeno, R. Scotti, F. Morazzoni, J. Arbiol, A. Cornet, J.R. Morante, NH_3 interaction with chromium-doped WO_3 nanocrystalline powders for gas sensing applications. J. Mater. Chem. **14**, 2412–2420 (2004)
19. E. Monti, G. Perletti, F. Piccinini, F. Monzini, F. Morazzoni, Interaction of Cu(II) and Cu (II)-anthracycline complexes with proteinkinase C. Spectromagnetic assessment of the inibitory effect. Inorganica Chimica Acta **205**, 181–184 (1993)
20. E. Bolzacchini, C. Canevali, F. Morazzoni, M. Orlandi, B. Rindone, R. Scotti, Spectromagnetic investigation of the active species in the oxidation of propenoidic phenols catalyzed by [NN′-bis(salicylidene)-ethane-1,2-diaminato)]cobalt(II) J. Chem. Soc. Dalton Trans. 4695–4699 (1997)
21. N. Chiodini, F. Morazzoni, A. Paleari, R. Scotti, G. Spinolo, Sol-gel synthesis of monolithic tin-doped silica glass. J. Mater. Chem. **9**, 2653–2658 (1999)
22. R. Scotti, M. D'Arienzo, A. Testino, F. Morazzoni, Photocatalytic mineralization of phenol catalyzed by pure and mixed phase hydrothermal titanium dioxide. Appl. Catal. B Environ. **88**, 497–504 (2009)
23. R. Scotti, I.R. Bellobono, C. Canevali, C. Cannas, M. Catti, M. D'Arienzo, A. Musinu, S. Polizzi, M. Sommariva, A. Testino, F. Morazzoni, Sol-gel pure and mixed-phase titanium dioxide for photocatalytic purposes: relations between phase composition, catalytic activity, and charge—trapped sites. Chem. Mater. **20**, 4051–4061 (2008)
24. M. D'Arienzo, J. Carbajo, A. Bahamonde, M. Crippa, S. Polizzi, R. Scotti, L. Wahba, F. Morazzoni, Photogenerated defects in shape-controlled TiO_2 anatase nanocrystals: A probe to evaluate the role of crystal facets in potocatalytic processs. J. Am. Chem. Soc. **133**, 17652–17661 (2011)
25. M. D'Arienzo, M. Crippa, A.A. Essawy, R. Scotti, L. Wahba, F. Morazzoni, P. Gentile, I.R. Bellobono, S. Polizzi, Membrane-assisted charge separation and photocatalytic activity in embedded TiO_2. A kinetic and mechanistic study. J. Phys. Chem. C **114**, 15755–15762 (2010)
26. M. D'Arienzo, D. Cristofori, R. Scotti, F. Morazzoni, New Insights into the SnO_2 Sensing Mechanism Based on the Properties of Shape Controlled Tin Oxide Nanoparticles. Chem. Mater. **25**, 3675–3686 (2013)
27. R. Scotti, L. Wahba, M. Crippa, M. D'Arienzo, R. Donetti, N. Santo, F. Morazzoni, Rubber-silica nanocomposites obtained by *in situ* sol-gel method: Particle shape influence on the filler-filler and filler-rubber interactions. Soft Matter. **8**, 2131 (2012)
28. L. Tadiello, M. D'Arienzo, B. Di Credico, T. Hanel, L. Matejka, M. Mauri, F. Morazzoni, R. Simonutti, M. Spirkova, R. Scotti, The filler-rubber interface in styrene butadiene nanocomposites with anisotropic silica particles: Morphology and dynamic properties. Soft Matter **11**, 4022–4033 (2015)
29. Q.B. Lu, S. Kalantari, C.R. Wang, Electron transfer reaction mechanism of cisplatin with DNA at the molecular level. Mol. Pharm. **4**, 624–628 (2007)
30. G. Zerbi, A. Barbon, R. Bengalli, A. Lucotti, T. Catelani, F. Tampieri, M. Gualtieri, M. D'Arienzo, F. Morazzoni, M. Camatini, Graphite particles induce ROS formation in cell free systems and human cell. Nanoscale **9**, 13640–13650 (2017)

31. C. Greco, U. Cosentino, D. Pitea, G. Moro, S. Santangelo, S. Patané, M. D'Arienzo, M. Fiore, F. Morazzoni, R. Ruffo, Role of the carbon defects in the catalytic oxygen reduction by graphite nanoparticles: A spectromagnetic, electrochemical and computational integrated approach. Phys. Chem. Chem. Phys. **21**, 6021–6032 (2019)
32. M. D'Arienzo, L. Gamba, F. Morazzoni, U. Cosentino, C. Greco, M. Lasagni, D. Pitea, G. Moro, C. Cepek, V. Butera, E. Sicilia, N. Russo, A.B. Munoz-Garcia, M. Pavone, exprimental and theoretical investigation on the catalytic generation of environmentally persistent free radicals from benzene. J.Phys.Chem. C **121**, 9381–9393 (2017)